RNA Processing

Volume I

A Practical Approach

Edited by
STEPHEN J. HIGGINS
and
B. DAVID HAMES
Department of Biochemistry and Molecular Biology, University of Leeds

—at—
OXFORD UNIVERSITY PRESS
Oxford New York Tokyo

The Practical Approach Series

SERIES EDITORS

D. RICKWOOD
*Department of Biology, University of Essex
Wivenhoe Park, Colchester, Essex CO4 3SQ, UK*

B. D. HAMES
*Department of Biochemistry and Molecular Biology
University of Leeds, Leeds LS2 9JT, UK*

Affinity Chromatography
Anaerobic Microbiology
Animal Cell Culture (2nd edition)
Animal Virus Pathogenesis
Antibodies I and II
Behavioural Neuroscience
Biochemical Toxicology
Biological Data Analysis
Biological Membranes
Biomechanics—Materials
Biomechanics—Structures and Systems
Biosensors
Carbohydrate Analysis
Cell–Cell Interactions
The Cell Cycle
Cell Growth and Division
Cellular Calcium
Cellular Interactions in Development
Cellular Neurobiology
Centrifugation (2nd edition)
Clinical Immunology
Computers in Microbiology
Crystallization of Nucleic Acids and Proteins
Cytokines
The Cytoskeleton
Diagnostic Molecular Pathology I and II
Directed Mutagenesis

DNA Cloning I, II, and III
Drosophila
Electron Microscopy in Biology
Electron Microscopy in Molecular Biology
Electrophysiology
Enzyme Assays
Essential Developmental Biology
Essential Molecular Biology I and II
Experimental Neuroanatomy
Fermentation
Flow Cytometry
Gas Chromatography
Gel Electrophoresis of Nucleic Acids (2nd edition)
Gel Electrophoresis of Proteins (2nd edition)
Gene Targeting
Gene Transcription
Genome Analysis
Glycobiology
Growth Factors
Haemopoiesis
Histocompatibility Testing
HPLC of Macromolecules
HPLC of Small Molecules
Human Cytogenetics I and II (2nd edition)
Human Genetic Disease Analysis

Immobilised Cells and Enzymes
Immunocytochemistry
In Situ Hybridization
Iodinated Density Gradient Media
Light Microscopy in Biology
Lipid Analysis
Lipid Modification of Proteins
Lipoprotein Analysis
Liposomes
Lymphocytes
Mammalian Cell Biotechnology
Mammalian Development
Medical Bacteriology
Medical Mycology
Microcomputers in Biochemistry
Microcomputers in Biology
Microcomputers in Physiology
Mitochondria
Molecular Genetic Analysis of Populations
Molecular Imaging in Neuroscience
Molecular Neurobiology
Molecular Plant Pathology I and II
Molecular Virology
Monitoring Neuronal Activity
Mutagenicity Testing
Neural Transplantation
Neurochemistry
Neuronal Cell Lines
NMR of Biological Macromolecules
Nucleic Acid and Protein Sequence Analysis
Nucleic Acid Hybridisation
Nucleic Acids Sequencing
Oligonucleotides and Analogues
Oligonucleotide Synthesis
PCR
Peptide Hormone Action
Peptide Hormone Secretion
Photosynthesis: Energy Transduction
Plant Cell Culture
Plant Molecular Biology
Plasmids (2nd edition)
Pollination Ecology
Post-implantation Mammalian Embryos
Preparative Centrifugation
Prostaglandins and Related Substances
Protein Architecture
Protein Engineering
Protein Function
Protein Phosphorylation
Protein Purification Applications
Protein Purification Methods
Protein Sequencing
Protein Structure
Protein Targeting
Proteolytic Enzymes
Radioisotopes in Biology
Receptor Biochemistry
Receptor–Effector Coupling
Receptor–Ligand Interactions
Ribosomes and Protein Synthesis
RNA Processing
Signal Transduction
Solid Phase Peptide Synthesis
Spectrophotometry and Spectrofluorimetry
Steroid Hormones
Teratocarcinomas and Embryonic Stem Cells
Transcription Factors
Transcription and Translation
Tumour Immunobiology
Virology
Yeast

Oxford University Press, Walton Street, Oxford OX2 6DP

Oxford New York Toronto
Delhi Bombay Calcutta Madras Karachi
Petaling Jaya Singapore Hong Kong Tokyo
Nairobi Dar es Salaam Cape Town
Melbourne Auckland

and associated companies in
Berlin Ibadan

Oxford is a trade mark of Oxford University Press

A Practical Approach 🛆 is a registered trade mark
of the Chancellor, Masters, and Scholars of the University of Oxford
trading as Oxford University Press

Published in the United States
by Oxford University Press, New York

© Oxford University Press, 1994

All rights reserved. No part of this publication may be
reproduced, stored in a retrieval system, or transmitted, in any
form or by any means, without the prior permission in writing of Oxford
University Press. Within the UK, exceptions are allowed in respect of any
fair dealing for the purpose of research or private study, or criticism or
review, as permitted under the Copyright, Designs and Patents Act, 1988, or
in the case of reprographic reproduction in accordance with the terms of
licences issued by the Copyright Licensing Agency. Enquiries concerning
reproduction outside those terms and in other countries should be sent to
the Rights Department, Oxford University Press, at the address above.

This book is sold subject to the condition that it shall not,
by way of trade or otherwise, be lent, re-sold, hired out, or otherwise
circulated without the publisher's prior consent in any form of binding
or cover other than that in which it is published and without a similar
condition including this condition being imposed
on the subsequent purchaser.

Users of books in the Practical Approach Series are advised that prudent
laboratory safety procedures should be followed at all times. Oxford
University Press makes no representation, express or implied, in respect of
the accuracy of the material set forth in books in this series and cannot
accept any legal responsibility or liability for any errors or omissions
that may be made.

A catalogue record for this book is available from the British Library

Library of Congress Cataloging in Publication Data
RNA processing: a practical approach / edited by Stephen J. Higgins
and B. David Hames.
    p.   cm.—(The Practical approach series ; 135, 136)
Includes bibliographical references and index.
    1. RNA—Research—Laboratory manuals. 2. Genetic regulation—
Research—Laboratory manuals. I. Higgins, S. J. (Steve J.)
II. Hames, B. D. III. Series.
QP623.R58   1993   574.87'3283—dc20   93-23019

ISBN 0 19 963344 4 (h/b) ⎫ Volume I
ISBN 0 19 963343 6 (p/b) ⎭

ISBN 0 19 963471 8 (h/b) ⎫ Volume II
ISBN 0 19 963470 X (p/b) ⎭

Typeset by Dobbie Typesetting Limited, Tavistock, Devon
Printed and bound in Great Britain by Information Press, Ltd., Eynsham, Oxon

Foreword

RNA processing: two helpful guides for cutting, pasting, trimming, and editing RNA

TOM MANIATIS

During the past decade, the study of RNA processing has risen to the forefront of molecular biology. At one time, the RNA processing field was a relatively small group of investigators interested in the trimming and packaging of prokaryotic transfer RNA and ribosomal RNAs. However, with the discovery of a bewildering array of RNA processing events in both prokaryotes and eukaryotes, the field has become a burgeoning enterprise. This change is reflected in the annual RNA processing meeting which has been rapidly transformed from a small informal gathering to one bursting the seams of the expanded meeting facilities at the Cold Spring Harbor Laboratory. Due to the pace of this unprecedented growth in size and diversity, no one has stopped long enough to compile a detailed description of even the most basic techniques used to study RNA. Students in the field are therefore introduced to the laboratory with handwritten protocols passed from person to person. The methods compiled in this book and its companion volume by leaders in the field should therefore contribute significantly to the training of new students and hopefully lead to the development of new techniques.

Volumes I and II of *RNA processing: a practical approach* cover virtually all aspects of RNA processing, including capping, splicing of both pre-mRNA and tRNA, polyadenylation, editing and ribosomal RNA processing. In addition, an overview of ribozymes is provided. An understanding of the various types of RNA processing *in vivo* is an essential prerequisite for studying *in vitro* the mechanisms involved. Therefore, the description of methods for RNA mapping, including nuclease S1 mapping, primer extension analysis, and the use of the polymerase chain reaction should be very useful.

An important technical advance in the analysis of RNA processing *in vitro* was the development of plasmid vectors containing bacteriophage-specific promoters for synthesizing labelled substrate RNAs. A description of the synthesis and purification of *in vitro* transcripts produced with SP6, T7, and T3 polymerase is therefore an essential part of the coverage. Also important is a description of the preparation and optimization of nuclear extracts for several types of *in vitro* RNA processing, 3'-end formation and polyadenylation. The manipulation and fractionation of these extracts have led to the purification of individual processing components and the cloning and characterization of the corresponding genes. Methods have also been developed for the selective inactivation of individual snRNPs using specific DNA oligonucleotides, or the depletion or purification of snRNP particles using oligo(2'-O-alkylribonucleotides). In addition, immunoaffinity purification methods for purifying snRNPs are presented as well as procedures for reconstituting individual snRNPs from purified proteins and RNAs. These tools for characterizing snRNPs will become increasingly important as the details of protein–protein and protein–RNA interactions in the spliceosome are unravelled.

An area less intensely investigated but of increasing importance, is the turnover of mRNA. There are now many examples in which gene expression during development is regulated by the selective turnover of specific RNAs. In addition, it is now clear that the rapid decrease in the expression of cytokines after induction by extracellular inducers involves a specific recognition sequence in the 3' non-coding sequence of mRNA. The chapter describing methods for studying mRNA turnover *in vitro* will stimulate further studies of this important problem.

The most recently discovered and most unexpected forms of RNA processing are autocatalytic splicing and RNA editing. The discovery of group I intron self-splicing and the RNase P ribozyme has led rapidly to detailed characterization of the catalytic reactions involved. Dissection and manipulation of the group I intron ribozyme have led to the creation of new catalytic activities and an understanding of the role of RNA secondary and tertiary structure in ribozyme function. Similarly, the elucidation of the structure and activity of group II introns has provided important new insights into the nature of ribozymes and has led to critical insights into the mechanisms involved in pre-mRNA splicing. Both group II intron and pre-mRNA splicing proceed through a similar branched RNA intermediate. Although the former process is autocatalytic and the latter requires multiple components assembled into a spliceosome, the catalytic events may be quite similar. In fact, the parallels between the role of RNA structure in group II intron splicing and the role of snRNA-pre-mRNA interactions have increased by the recent demonstration of dynamic interactions between snRNA and specific sequences in pre-mRNA during the pre-mRNA splicing reaction. The simple view is that the same catalytic mechanisms are involved in both processes. However, in the case of pre-mRNA processing the RNA–RNA interactions are mediated by spliceosomal proteins. More practical applications of ribozymology are emerging from the study of hammerhead and hairpin ribozymes, another topic discussed in the ribozyme chapter. The manipulation of these interesting molecules and their use in targeting and processing specific RNA transcripts are leading to a better understanding of RNA catalysis.

RNA editing, the process of post-transcriptional insertion, deletion or substitution of specific bases in mRNA, is the most puzzling example of RNA

Foreword

processing. The discovery of guide RNAs complementary to edited portions of mRNA and the identification of putative intermediates in the editing processes have led to the proposal of a specific mechanism for RNA editing. This mechanism involves an orderly cycling of the editing process in a 3' to 5' direction. Specific catalytic mechanisms have been proposed based in part on comparisons to the group I and II self-splicing reactions. The detailed description in *RNA processing: a practical approach* of the biological systems and techniques used to study RNA editing should stimulate the development of new approaches to the study of the mechanisms involved.

Although most of this two volume set focuses on the use of biochemical approaches to RNA processing, a description of genetic techniques used to study pre-mRNA processing in yeast is also provided. The application of these techniques has led to the identification and cloning of genes encoding essential splicing factors (PRP genes). In addition, a variety of genetic tools, including gene 'knockouts', targeted mutagenesis, interactive suppression, and conditional expression vectors have been used to study the function of a number of yeast splicing components including PRP proteins and specific snRNAs. These tools have been used in conjunction with *in vitro* studies using nuclear extracts prepared from wild type yeast strains as well as strains lacking an essential splicing factor. Comparison of the results of these studies with those of mammalian splicing has revealed a remarkable conservation in the mechanisms of the splicing reaction and spliceosome assembly, and in the role of snRNPs in splice site recognition. This comparison of yeast and mammalian splicing points to an additional benefit of collecting a variety of RNA processing methods together. It provides the opportunity to examine the similarities and differences in experimental approaches established for studying different types of RNA processing. Hopefully, this will lead to previously unrecognized connections and new ideas.

Preface

This book arose out of the success of a book we edited for the Practical Approach series in 1984, entitled *Transcription and translation: a practical approach*. When the time came to consider organizing a second edition, it rapidly became clear that no one book of the desired size could include, in sufficient detail, the myriad of important new techniques, particularly in the area of post-transcriptional processing of RNA transcripts, that had arisen since the first edition. Thus the logical decision was taken to produce several books, the first of which, *Gene transcription: a practical approach*, has recently been published. *RNA processing: a practical approach* Volumes I and II are the next books in this planned set and are companion volumes.

RNA processing Volume I begins with Benoit Chabot describing the synthesis and purification of RNA substrates for RNA processing investigations. Paula Grabowski then covers essential methods for the identification and analysis of spliced mRNAs. Next, Ian Eperon and Adrian Krainer have collaborated to produce an important chapter on the analysis of the splicing of mRNA precursors in mammalian cells. Another key area, the analysis of RNP complexes and their interactions, is covered in two chapters by Angus Lamond and Brian Sproat and by Reinhard Lührmann and his colleagues. These are all central issues in RNA processing studies. Finally, Andrew Newman has contributed a chapter on investigations of pre-mRNA splicing in yeast, which allows some approaches that are not possible with mammalian cells, most notably the isolation and analysis of splicing mutants.

The companion volume, *RNA processing* Volume II, includes contributions from Walter Keller's laboratory on techniques for analysis of 3' end-processing of mRNA and from Aaron Shatkin's group on capping and methylation of mRNA. RNA editing and the analysis mRNA turnover are also included in this volume, with chapters by Larry Simpson and his colleagues and Jeff Ross, respectively. Barbara Sollner-Webb and Cathy Enright describe methods for studying ribosomal RNA processing and Chris Greer similarly covers transfer RNA processing. Finally, David Shub, Craig Peebles, and Arnold Hampel have combined their efforts to produce a very topical chapter covering investigations of detailed splicing reactions, comprehensively covering group I intron slicing, group II intron splicing and hammerhead and hairpin ribozymes.

The aim of both books remains the aim of this popular series; to present, in a clear readable manner, the background to the range of techniques and experimental approaches available, to describe in precise detail a key selection of tried and tested protocols, and to discuss potential pitfalls, data interpretation, and a variety of other hints and tips for the active scientist. It is a measure of the efforts of our contributors that we believe these aims have been more than

Preface

met in each volume. We thank them for their diligence in writing texts which address these aims and, where we felt that editorial changes were essential, for graciously accepting these. We hope and believe that the end result will be seen as a comprehensive and valuable two volume compendium of the best of current methodology in this subject area. Because of the scientific quality of the contributions and the deliberately explanatory style of writing, we are confident that, like their predecessor, these books will rightfully enjoy popularity among both newcomers to the field and more experienced researchers.

Leeds   Stephen Higgins
February 1993   David Hames

Contents

List of contributors	xviii
Abbreviations	xx

1. Synthesis and purification of RNA substrates   1
Benoit Chabot

    1. Introduction   1
        Creating an RNase-free environment   1

    2. Synthesis of RNA substrates   2
        Preparation of template DNA   3
        Transcription using SP6, T3, and T7 RNA polymerases   8
        Purification of *in vitro* RNA transcripts   14

    3. Purification of RNA from cells   21
        Purification of total RNA   21
        Purification of cytoplasmic RNA   24
        Purification of poly(A)$^+$ RNA   25

    4. Purification of snRNAs   27

    5. Purification of other RNAs   28
        Acknowledgements   28
        References   28

2. Characterization of RNA   31
Paula J. Grabowski

    1. Introduction   31

    2. Isolation of RNA   31

    3. Procedures for establishing the identity of spliced RNAs   32
        Introduction   32
        Basic nuclease S1 analysis   32
        Combined RNase H and nuclease S1 analysis   38
        RT–PCR analysis   40
        Primer extension analysis of RNA   43
        Northern blot analysis with exon and intron probes   46

Contents

4. Mapping of RNA branchpoints and 5' ends of introns   49
    Introduction   49
    Mapping RNA branchpoints by primer extension   50
    Mapping the 5' end of an intron   53

Acknowledgements   55

References   55

3. Splicing of mRNA precursors in mammalian cells   57
Ian C. Eperon and Adrian R. Krainer

1. Introduction   57

2. Analysis of pre-mRNA splicing in mammalian cells   58
    Expression of exogenous genes   58
    Analysis of RNA expressed   61

3. Analysis of RNA splicing in metazoan nuclear extracts   73
    Preparation of active extracts   73
    Splicing reactions *in vitro*   75
    Depletion of components from nuclear extracts   82

4. Fractionation of splicing extracts and biochemical identification of splicing factors   83
    Biochemical fractionation strategies   83
    Purification procedures for specific splicing factors   90
    Isolation of cDNAs for splicing factors and expression of the recombinant proteins   98

Acknowledgements   98

References   98

4. Isolation and characterization of ribonucleoprotein complexes   103
Angus I. Lamond and Brian S. Sproat

1. Introduction   103

2. Chemical synthesis of oligoribonucleotides and oligo(2'-O-alkylribonucleotides)   104
    Advantages of using antisense oligonucleotides for studying RNP structure and function   104
    General points concerning the chemical synthesis of oligoribonucleotides   104
    Chemical synthesis of oligoribonucleotides   106
    Chemical synthesis of oligo(2'-*O*-methylribonucleotides) and oligo-(2'-*O*-allylribonucleotides)   108

Contents

3. Antisense affinity selection of RNAs or RNP complexes 110
 Strategy 110
 Probes for affinity selection 111
 Biotin–streptavidin affinity selection of RNP complexes 113

4. Depletion of cell extracts of RNAs or RNP complexes using affinity selection with antisense oligonucleotide probes 117

5. Analysis of proteins in RNP complexes purified by affinity selection with oligonucleotide probes 120
 Introduction 120
 Antisense affinity selection of RNP complexes from precleared cell extracts 120

6. Targeted RNase H cleavage of RNA 123

7. Analysis of RNAs and RNPs by gel electrophoresis 126
 Introduction 126
 RNA separation by denaturing polyacrylamide gel electrophoresis 127
 Separation of RNP complexes by electrophoresis in native polyacrylamide–agarose gels 128
 Transfer of RNAs from gels by electroblotting 130
 Analysis of RNAs by Northern hybridization with riboprobes 131

8. Chemical modification interference analysis of RNA 134
 Strategy 134
 3′ End-labelling of RNA using RNA ligase 135
 Chemical modification of RNA at purine and pyrimidine bases 136
 Cleavage of modified RNA with aniline 138

References 139

5. Analysis of ribonucleoprotein interactions 141
Cindy L. Will, Berthold Kastner, and Reinhard Lührmann

1. Introduction 141

2. Immunoaffinity purification of RNP complexes 142
 Strategy 142
 Immobilization of antibodies on Sepharose matrices 142
 Immunoaffinity chromatography of snRNPs 145

3. Ion-exchange chromatography of snRNPs 149

4. Analysis of snRNP protein and RNA by gel electrophoresis 151

5. Reconstitution of snRNPs from purified RNA and protein 156
 Introduction 156
 Preparation of snRNP protein 156
 RNA preparation and snRNP reconstitution 158

Contents

6.	Analysis of protein–RNA interactions by immunoprecipitation of snRNPs	159
	Strategy	159
	Immunoprecipitation of snRNPs	160
	Analysis of immunoprecipitated snRNPs	162
7.	Analysis of RNA–protein interactions by UV cross-linking	163
	Strategy	163
	snRNP preparation and UV irradiation	164
	Analysis of the extent of UV cross-linking	165
	Identification of cross-linked proteins	165
	Identification of RNA cross-link sites	167
8.	Mapping of RNA–protein interactions by RNase protection	168
9.	Immunoelectron microscopy of snRNP particles	170
	Introduction	170
	Stabilization of snRNP–protein interactions for immunoelectron microscopy	170
	Formation of antibody–snRNP complexes	171
	Preparation of samples for electron microscopy	174
	Investigation of snRNPs and snRNP–IgG complexes by electron microscopy	174

Acknowledgements   176

References   176

6. Analysis of pre-mRNA splicing in yeast   179
Andrew Newman

1.	Introduction	179
2.	Expression of exogenous genes	179
	Vectors	179
	Transformation of yeast	180
3.	Analysis of yeast RNA	182
	Isolation of RNA from yeast	182
	RNA analysis by primer extension	184
4.	Genetic strategies for investigating pre-mRNA splicing	186
	Isolation and characterization of *PRP* genes and their products	186
	Genetic approaches to splicing based on interactive suppression	187
	Functional analysis of splicing factors *in vivo*	188
5.	Analysis of RNA splicing in yeast extracts	190
	Preparation of active extracts from yeast	190
	Splicing reactions *in vitro*	191
	Depletion of components from splicing extracts	193

References   194

Appendix: Suppliers of specialist items	197
Contents of Volume II	203
Index	205

Contributors

BENOIT CHABOT
Département de Microbiologie, Faculté de Médecine, 3001, 12th avenue North, Université de Sherbrooke, Sherbrooke, Québec, Canada J1H 5N4.

IAN C. EPERON
Department of Biochemistry, University of Leicester, Adrian Building, University Road, Leicester LE1 7RH.

PAULA J. GRABOWSKI
Department of Biological Sciences, University of Pittsburgh, Pittsburgh, PA 15260, USA.

B. DAVID HAMES
Department of Biochemistry and Molecular Biology, University of Leeds, Leeds LS2 9JT.

STEPHEN J. HIGGINS
Department of Biochemistry and Molecular Biology, University of Leeds, Leeds LS2 9JT.

BERTHOLD KASTNER
Institut für Molekularbiologie und Tumorforschung, Klinikum der Philipps-Universität Marburg, D-35037 Marburg, Germany.

ADRIAN R. KRAINER
Cold Spring Harbor Laboratory, Box 100, 1 Bungtown Road, Cold Spring Harbor, NY 11724-2208, USA.

ANGUS I. LAMOND
Gene Expression Programme, European Molecular Biology Laboratory, Postfach 10.2209, Meyerhofstrasse 1, D-6900 Heidelberg, Germany.

REINHARD LÜHRMANN
Institut für Molekularbiologie und Tumorforschung, Klinikum der Philipps-Universität Marburg, Emil Mannkopff Straße 2, D-35037 Marburg, Germany.

ANDREW NEWMAN
Laboratory of Molecular Biology, MRC Centre, Hills Road, Cambridge CB2 2QH, UK.

Contributors

BRIAN S. SPROAT
Biochemical Instrumentation Programme, European Molecular Biology Laboratory, Postfach 10.2209, Meyerhofstrasse 1, D-6900 Heidelberg, Germany.

CINDY L. WILL
Institut für Molekularbiologie und Tumorforschung, Klinikum der Philipps-Universität Marburg, Emil Mannkopff Straße 2, D-35037 Marburg, Germany.

Abbreviations

A	adenine/adenosine/adenylate
A_{250}	absorption at 250 nm
A_{600}	absorption at 600 nm
ADH	alcohol dehydrogenase
AMV	avian myeloblastosis virus
ATP	adenosine 5'-triphosphate
bp	base pair(s)
BSA	bovine serum albumin
C	cytosine/cytidine/cytidylate
cDNA	complementary DNA
cDNA-PCR	complementary DNA-polymerase chain reaction
CNBr	cyanogen bromide
c.p.m.	counts per minute
CTP	cytidine 5'-triphosphate
Da	dalton(s)
dATP	deoxyadenosine 5'-triphosphate
dCTP	deoxycytidine 5'-triphosphate
ddNTP	2',3'-dideoxyribonucleotide 5'-triphosphate
DEAE	diethyl aminoethyl
DEPC	diethyl pyrocarbonate
dGTP	deoxyguanosine 5'-triphosphate
DMEM	Dulbecco's modified Eagle's medium
DMSO	dimethyl sulphoxide
DNase	deoxyribonuclease
dNTP	deoxyribonucleoside 5'-triphosphate
d.p.m.	disintegrations per minute
DSP	dithio*bis*-succinimidyl propionate
DTE	dithioerythritol
DTT	dithiothreitol
dTTP	deoxythymidine 5'-triphosphate
EDTA	ethylenediamine-tetraacetic acid
EGTA	ethylene glycol-*O*,*O'*-*bis*(2-aminoethyl)-*N*,*N*,*N'*,*N'*-tetraacetic acid
ELISA	enzyme-linked immunosorbent assay
5-FOA	5-fluoroorotic acid
FPLC	fast-performance liquid chromatography
G	guanine/guanosine/guanylate
g	gravity
GTP	guanosine 5'-triphosphate

Abbreviations

h	hour(s)
HBS	Hepes-buffered saline
HPLC	high-performance liquid chromatography
IgG	immunoglobulin class G
IgM	immunoglobulin class M
IVS	intervening sequence (intron)
kb(p)	kilobase (pair)
kDa	kilodaltons
kV	kilovolts
m^6A	6-methyl adenine
m_3G	2,2,7-trimethyl guanosine
m^7G	7-methyl guanine
min	minute(s)
mJ	millijoules
M_r	relative molecular weight
mRNA	messenger RNA
NEO	neomycin
nt	nucleotide
NTP	nucleoside 5'-triphosphate
PAGE	polyacrylamide gel electrophoresis
PAS	Protein A–Sepharose
PBS	phosphate-buffered saline
PCR	polymerase chain reaction
PCV	packed cell volume
PEG	polyethylene glycol
PMSF	phenylmethylsulphonyl fluoride
PNK	polynucleotide kinase
p.p.m.	parts per million
PVA	polyvinyl alcohol
Poly(A)$^+$ RNA	polyadenylated RNA
rATP	adenosine 5'-triphosphate
rCTP	cytidine 5'-triphosphate
rGTP	guanosine 5'-triphosphate
RNase	ribonuclease
RNP	ribonucleoprotein (particle)
rNTP	ribonucleoside 5'-triphosphate
r.p.m.	revolutions per minute
rRNA	ribosomal RNA
RT	reverse transcriptase
RT–PCR	reverse transcription–polymerase chain reaction
rUTP	uridine 5'-triphosphate
S	Svedberg
SDS	sodium dodecyl sulphate
SDS–PAGE	SDS–polyacrylamide gel electrophoresis

Abbreviations

snRNA	small nuclear RNA
snRNP	small nuclear ribonucleoprotein (particle)
SSC	standard saline citrate
TAE	Tris–acetate–EDTA buffer
TBE	Tris–borate–EDTA buffer
TE	Tris–EDTA buffer
TEMED	N,N,N',N'-tetramethylenediamine
T_m	melting temperature
tRNA	transfer RNA
TTP	(deoxy)thymidine 5′-triphosphate
UTP	uridine 5′-triphosphate
UV	ultraviolet
V	volts
X-gal	5-bromo-4-chloro-3-indolyl-β-D-galactoside
YEPD	yeast extract, peptone, dextrose

Contents

List of contributors	xviii
Abbreviations	xx

1. Synthesis and purification of RNA substrates   1
Benoit Chabot

   1. Introduction   1
       Creating an RNase-free environment   1

   2. Synthesis of RNA substrates   2
       Preparation of template DNA   3
       Transcription using SP6, T3, and T7 RNA polymerases   8
       Purification of *in vitro* RNA transcripts   14

   3. Purification of RNA from cells   21
       Purification of total RNA   21
       Purification of cytoplasmic RNA   24
       Purification of poly(A)$^+$ RNA   25

   4. Purification of snRNAs   27

   5. Purification of other RNAs   28

       Acknowledgements   28
       References   28

2. Characterization of RNA   31
Paula J. Grabowski

   1. Introduction   31

   2. Isolation of RNA   31

   3. Procedures for establishing the identity of spliced RNAs   32
       Introduction   32
       Basic nuclease S1 analysis   32
       Combined RNase H and nuclease S1 analysis   38
       RT–PCR analysis   40
       Primer extension analysis of RNA   43
       Northern blot analysis with exon and intron probes   46

Contents

4. Mapping of RNA branchpoints and 5′ ends of introns   49
    Introduction   49
    Mapping RNA branchpoints by primer extension   50
    Mapping the 5′ end of an intron   53

Acknowledgements   55

References   55

3. Splicing of mRNA precursors in mammalian cells   57
Ian C. Eperon and Adrian R. Krainer

1. Introduction   57

2. Analysis of pre-mRNA splicing in mammalian cells   58
    Expression of exogenous genes   58
    Analysis of RNA expressed   61

3. Analysis of RNA splicing in metazoan nuclear extracts   73
    Preparation of active extracts   73
    Splicing reactions *in vitro*   75
    Depletion of components from nuclear extracts   82

4. Fractionation of splicing extracts and biochemical identification of splicing factors   83
    Biochemical fractionation strategies   83
    Purification procedures for specific splicing factors   90
    Isolation of cDNAs for splicing factors and expression of the recombinant proteins   98

Acknowledgements   98

References   98

4. Isolation and characterization of ribonucleoprotein complexes   103
Angus I. Lamond and Brian S. Sproat

1. Introduction   103

2. Chemical synthesis of oligoribonucleotides and oligo(2′-O-alkylribonucleotides)   104
    Advantages of using antisense oligonucleotides for studying RNP structure and function   104
    General points concerning the chemical synthesis of oligoribonucleotides   104
    Chemical synthesis of oligoribonucleotides   106
    Chemical synthesis of oligo(2′-O-methylribonucleotides) and oligo-(2′-O-allylribonucleotides)   108

3. Antisense affinity selection of RNAs or RNP complexes 110
   Strategy 110
   Probes for affinity selection 111
   Biotin–streptavidin affinity selection of RNP complexes 113

4. Depletion of cell extracts of RNAs or RNP complexes using affinity selection with antisense oligonucleotide probes 117

5. Analysis of proteins in RNP complexes purified by affinity selection with oligonucleotide probes 120
   Introduction 120
   Antisense affinity selection of RNP complexes from precleared cell extracts 120

6. Targeted RNase H cleavage of RNA 123

7. Analysis of RNAs and RNPs by gel electrophoresis 126
   Introduction 126
   RNA separation by denaturing polyacrylamide gel electrophoresis 127
   Separation of RNP complexes by electrophoresis in native polyacrylamide–agarose gels 128
   Transfer of RNAs from gels by electroblotting 130
   Analysis of RNAs by Northern hybridization with riboprobes 131

8. Chemical modification interference analysis of RNA 134
   Strategy 134
   3' End-labelling of RNA using RNA ligase 135
   Chemical modification of RNA at purine and pyrimidine bases 136
   Cleavage of modified RNA with aniline 138

References 139

5. Analysis of ribonucleoprotein interactions 141
Cindy L. Will, Berthold Kastner, and Reinhard Lührmann

1. Introduction 141

2. Immunoaffinity purification of RNP complexes 142
   Strategy 142
   Immobilization of antibodies on Sepharose matrices 142
   Immunoaffinity chromatography of snRNPs 145

3. Ion-exchange chromatography of snRNPs 149

4. Analysis of snRNP protein and RNA by gel electrophoresis 151

5. Reconstitution of snRNPs from purified RNA and protein 156
   Introduction 156
   Preparation of snRNP protein 156
   RNA preparation and snRNP reconstitution 158

6.	**Analysis of protein–RNA interactions by immunoprecipitation of snRNPs**	159
	Strategy	159
	Immunoprecipitation of snRNPs	160
	Analysis of immunoprecipitated snRNPs	162
7.	**Analysis of RNA–protein interactions by UV cross-linking**	163
	Strategy	163
	snRNP preparation and UV irradiation	164
	Analysis of the extent of UV cross-linking	165
	Identification of cross-linked proteins	165
	Identification of RNA cross-link sites	167
8.	**Mapping of RNA–protein interactions by RNase protection**	168
9.	**Immunoelectron microscopy of snRNP particles**	170
	Introduction	170
	Stabilization of snRNP–protein interactions for immunoelectron microscopy	170
	Formation of antibody–snRNP complexes	171
	Preparation of samples for electron microscopy	174
	Investigation of snRNPs and snRNP–IgG complexes by electron microscopy	174
	Acknowledgements	176
	References	176

6. Analysis of pre-mRNA splicing in yeast — 179
Andrew Newman

1.	**Introduction**	179
2.	**Expression of exogenous genes**	179
	Vectors	179
	Transformation of yeast	180
3.	**Analysis of yeast RNA**	182
	Isolation of RNA from yeast	182
	RNA analysis by primer extension	184
4.	**Genetic strategies for investigating pre-mRNA splicing**	186
	Isolation and characterization of *PRP* genes and their products	186
	Genetic approaches to splicing based on interactive suppression	187
	Functional analysis of splicing factors *in vivo*	188
5.	**Analysis of RNA splicing in yeast extracts**	190
	Preparation of active extracts from yeast	190
	Splicing reactions *in vitro*	191
	Depletion of components from splicing extracts	193
	References	194

Appendix: Suppliers of specialist items	197
Contents of Volume II	203
Index	205

Contributors

BENOIT CHABOT
Département de Microbiologie, Faculté de Médecine, 3001, 12th avenue North, Université de Sherbrooke, Sherbrooke, Québec, Canada J1H 5N4.

IAN C. EPERON
Department of Biochemistry, University of Leicester, Adrian Building, University Road, Leicester LE1 7RH.

PAULA J. GRABOWSKI
Department of Biological Sciences, University of Pittsburgh, Pittsburgh, PA 15260, USA.

B. DAVID HAMES
Department of Biochemistry and Molecular Biology, University of Leeds, Leeds LS2 9JT.

STEPHEN J. HIGGINS
Department of Biochemistry and Molecular Biology, University of Leeds, Leeds LS2 9JT.

BERTHOLD KASTNER
Institut für Molekularbiologie und Tumorforschung, Klinikum der Philipps-Universität Marburg, D-35037 Marburg, Germany.

ADRIAN R. KRAINER
Cold Spring Harbor Laboratory, Box 100, 1 Bungtown Road, Cold Spring Harbor, NY 11724-2208, USA.

ANGUS I. LAMOND
Gene Expression Programme, European Molecular Biology Laboratory, Postfach 10.2209, Meyerhofstrasse 1, D-6900 Heidelberg, Germany.

REINHARD LÜHRMANN
Institut für Molekularbiologie und Tumorforschung, Klinikum der Philipps-Universität Marburg, Emil Mannkopff Straße 2, D-35037 Marburg, Germany.

ANDREW NEWMAN
Laboratory of Molecular Biology, MRC Centre, Hills Road, Cambridge CB2 2QH, UK.

Contributors

BRIAN S. SPROAT
Biochemical Instrumentation Programme, European Molecular Biology Laboratory, Postfach 10.2209, Meyerhofstrasse 1, D-6900 Heidelberg, Germany.

CINDY L. WILL
Institut für Molekularbiologie und Tumorforschung, Klinikum der Philipps-Universität Marburg, Emil Mannkopff Straße 2, D-35037 Marburg, Germany.

Abbreviations

A	adenine/adenosine/adenylate
A_{250}	absorption at 250 nm
A_{600}	absorption at 600 nm
ADH	alcohol dehydrogenase
AMV	avian myeloblastosis virus
ATP	adenosine 5'-triphosphate
bp	base pair(s)
BSA	bovine serum albumin
C	cytosine/cytidine/cytidylate
cDNA	complementary DNA
cDNA-PCR	complementary DNA-polymerase chain reaction
CNBr	cyanogen bromide
c.p.m.	counts per minute
CTP	cytidine 5'-triphosphate
Da	dalton(s)
dATP	deoxyadenosine 5'-triphosphate
dCTP	deoxycytidine 5'-triphosphate
ddNTP	2',3'-dideoxyribonucleotide 5'-triphosphate
DEAE	diethyl aminoethyl
DEPC	diethyl pyrocarbonate
dGTP	deoxyguanosine 5'-triphosphate
DMEM	Dulbecco's modified Eagle's medium
DMSO	dimethyl sulphoxide
DNase	deoxyribonuclease
dNTP	deoxyribonucleoside 5'-triphosphate
d.p.m.	disintegrations per minute
DSP	dithio*bis*-succinimidyl propionate
DTE	dithioerythritol
DTT	dithiothreitol
dTTP	deoxythymidine 5'-triphosphate
EDTA	ethylenediamine-tetraacetic acid
EGTA	ethylene glycol-O,O'-*bis*(2-aminoethyl)-N,N,N',N'-tetraacetic acid
ELISA	enzyme-linked immunosorbent assay
5-FOA	5-fluoroorotic acid
FPLC	fast-performance liquid chromatography
G	guanine/guanosine/guanylate
g	gravity
GTP	guanosine 5'-triphosphate

1
Synthesis and purification of RNA substrates

BENOIT CHABOT

1. Introduction

In recent years, the purification of RNA molecules from different organisms and cell types has been crucial to the analysis of gene expression and RNA processing events. Likewise, the rapid development of *in vitro* RNA maturation systems has relied extensively on the ability to synthesize and purify RNA substrates. Generating pure, undegraded RNA is thus critical to *in vivo* and *in vitro* analyses of RNA processing. For the most part, procedures used to obtain intact and pure RNA are relatively simple and have been described in detail in a variety of publications and technical books. This chapter does not attempt to provide a comprehensive coverage of all these procedures. Instead, it describes a few of the most rapid, convenient, and economical ways to prepare RNA for direct analysis or for use as substrates in *in vitro* systems of RNA processing.

1.1 Creating an RNase-free environment

Researchers who work with RNA are always very much concerned about RNase contamination and thus take great pains to establish an RNase-free micro-environment. However, in most cases problems with RNase contamination can be avoided by taking the following simple precautions:

(a) Autoclave plasticware and glassware.

(b) Wear disposable plastic or latex gloves to avoid RNase contamination from human skin.

(c) Sterilize all solutions by filtration through cellulose nitrate membranes (pore size 0.2 μm). This removes contaminating bacteria and their associated RNases. Cellulose nitrate membranes also have a general affinity for proteins and thus this step also removes soluble RNases. For large volumes (>25 ml), use disposable sterile filter units (Nalgene, 120-0020); for smaller volumes (down to 1 ml), use syringe filters (Nalgene, 190-2020). This method is an

1: Synthesis and purification of RNA substrates

easy and rapid alternative to autoclaving solutions. Ribonucleases (and in particular RNase A) are highly resistant to thermal and chemical stresses. In fact, destroying bacteria by autoclaving may actually release RNases into the solution. Filtration is also preferable to the common practice of treating solutions with diethylpyrocarbonate since this product is toxic and so must be handled with caution.

(d) Use one or more of the specific RNase inhibitors that are now available. Among these the placental RNase inhibitor is most effective against RNase A (pancreatic RNase) and is commercially available under various trade names, e.g. RNasin™ (Promega), RNAguard™ (Pharmacia), Inhibit-Ace™ (5 Prime-3 Prime, Inc.). However, this product is somewhat expensive, so it is usually reserved for incubations performed in small volumes (e.g. during *in vitro* transcription, RNA splicing, polyadenylation, etc.). A minimum concentration of dithiothreitol (DTT) (1 mM) is essential to maintain the inhibitory activity of this protein.

(e) Where possible, choose cells or tissues low in RNase for the source of the RNA to be purified.

(f) Use methods that involve rapid inactivation of endogenous RNases following cell disruption. Three widely-used procedures that make use of denaturing agents (urea and guanidinium thiocyanate) are described in Section 3.

(g) Rapid processing and the use of low temperatures (0–4 °C) are some of the best precautions one can take to guarantee the integrity of RNA preparations. The less an RNA sample is manipulated, the less likely it is to encounter nucleases. When manipulating RNA, minimize the time the RNA spends in an intermediate solution, especially if at room temperature. Have all your solutions ready and at hand when you begin an extraction procedure.

(h) Once pure RNA is obtained, store it at low temperatures (e.g. −70 °C), preferably in 70% ethanol.

2. Synthesis of RNA substrates

The controlled production of single-stranded RNA substrates of defined length from DNA templates using bacteriophage promoters and purified RNA polymerases has been, and continues to be, of fundamental importance for research on post-transcriptional RNA modifications. Moreover, this technology is now widely used to prepare templates for *in vitro* translation and to generate highly sensitive hybridization probes for Southern and Northern analyses, RNase protection mapping, and *in situ* hybridization. This section describes protocols to allow the experimenter to obtain undegraded RNA of high-specific activity for *in vitro* studies on RNA processing. Transcription protocols in this section are not significantly different from those already described elsewhere (1–3). However, they have been well-tested and work with any of the commonly-used purified bacteriophage RNA polymerases (T7, T3, or SP6).

The synthesis of RNA substrates involves three stages; preparation of the DNA template for transcription, the transcription reaction itself, and the purification of the RNA produced. Each of these is described in the following sections.

2.1 Preparation of template DNA

The usual method of *in vitro* transcription is to use a bacteriophage RNA polymerase to transcribe the template DNA inserted downstream from either an SP6, T7, or T3 phage promoter. Several plasmid vectors containing multiple cloning sites located between two such promoters are commercially available (e.g. the pKSBluescript, pT7T3, and pGEM series from Stratagene, Pharmacia, and Promega, respectively). By choosing the appropriate phage RNA polymerase one can produce large amounts of RNA, complementary to either strand of the template DNA.

To prepare the DNA template for the transcription reaction, the recombinant plasmid DNA must first be linearized by cleavage at an appropriate restriction enzyme site downstream of the promoter and the cloned insert. In the subsequent 'run-off' transcription reaction, the position of the cut will define the 3' end of the transcript.

Although the template DNA can be prepared by the rapid minipreparation method involving alkaline lysis (1), such DNA often transcribes poorly unless it is extensively cleaned. Therefore, to avoid this problem and for optimal results, prepare the plasmid DNA by caesium chloride (CsCl) gradient purification as described in *Protocol 1*.

Protocol 1. Preparation of template DNA from a recombinant plasmid

Equipment and reagents

- TB broth: Separately autoclave solution A (11.57 g KH_2PO_4 + 62.71 g K_2HPO_4 in 500 ml of H_2O) and solution B (60 g Bacto-tryptone, 120 g yeast extract, 20 ml glycerol in 4.5 litres of H_2O). TB broth is prepared by mixing 450 ml of solution B with 50 ml of solution A
- Sorvall centrifuge with GSA and SS34 rotors, Beckman ultracentrifuge with VTi65 rotor (or equivalents)
- Centrifuge bottles and tubes: 250 ml polypropylene (Nalgene, 312 0-0250), 50 ml polyallomer (Oak Ridge pattern; Nalgene, 3119-0050), 50 ml graduated (Falcon, 2070), 5 ml polyallomer (Beckman, Quick-Seal™) and sealing apparatus for centrifuge tubes (Tool kit for Quick-Seal™ tubes, Beckman)
- Solution I (0.9 g glucose, 2.5 ml 1 M Tris–HCl, pH 8.0, 2 ml 0.5 M EDTA, H_2O to 100 ml)
- Solution II (4 ml 5.0 M NaOH, 10 ml of 10% SDS, H_2O to 100 ml). Always use a fresh solution
- Solution III (29.45 g potassium acetate, 11.5 ml glacial acetic acid, H_2O to 100 ml)
- TE buffer (10 mM Tris–HCl, pH 8.0, 1 mM EDTA)
- Ethidium bromide (10 mg/ml H_2O)
- Butanol saturated with CsCl
- Appropriate restriction enzyme and buffer

Continued

1: Synthesis and purification of RNA substrates

Protocol 1. *Continued*

- Agarose gel prepared as described in Protocol 6
- Phenol:chloroform:isoamyl alcohol (25:24:1) saturated with 10 mM Tris–HCl, pH 7.5, 1 mM EDTA
- 3.0 M sodium acetate, pH 6.5

Method

1. Grow recombinant bacteria overnight in 200 ml of TB broth containing the appropriate antibiotic (e.g. ampicillin at 100 µg/ml).
2. Centrifuge the culture in a 250 ml polypropylene centrifuge bottle at 4 °C in a GSA rotor (Sorvall) at 3000 r.p.m. (1000 g) for 15 min.
3. Discard the supernatant, resuspend bacterial pellet in 7 ml of solution I and transfer this into a 50 ml polyallomer centrifuge tube (Oak ridge pattern).
4. Add 16 ml of solution II. Mix by inversion and let the mixture stand for 10 min at room temperature.
5. Add 12 ml of solution III. Mix by inversion and leave the mixture on ice for 1 h.
6. Centrifuge the solution for 30 min at 15 000 r.p.m. (17 500 g) in an SS34 rotor at 4 °C.
7. Divide the supernatant equally between two 50 ml graduated tubes (Falcon) and add an equal vol. of isopropanol to each tube.
8. Leave the mixture at −20 °C for at least 10 min.
9. Centrifuge the mixture for 10 min in a clinical centrifuge (600 g) at room temperature. Discard the supernatant.
10. Rinse the DNA pellets by resuspending in absolute ethanol. Pool the resuspended DNA and centrifuge it at 600 g. Allow the pellet to air dry.
11. Add 3 ml of TE buffer to dissolve the pellet. Measure the volume and add TE buffer to 4.2 ml final volume.
12. Add 5 g of CsCl powder and dissolve it by shaking.
13. Add 150 µl of ethidium bromide (10 mg/ml) and mix.
14. Transfer the solution into a 5 ml polyallomer Quick-Seal™ centrifuge tube.
15. Seal the tube and centrifuge at 50 000 r.p.m. (220 000 g) for at least 6 h in a VTi65 rotor.
16. Locate the plasmid DNA band under UV illumination and remove the DNA using a 1 ml syringe and an 18 g × 1½ inch needle[a]. Transfer the plasmid solution into a microcentrifuge tube.
17. Remove the ethidium bromide by extracting five times with 0.5 ml of butanol saturated with CsCl.
18. Remove the CsCl by dialysing the plasmid DNA solution overnight in 1 litre of H_2O at 4 °C.

Continued

Protocol 1. *Continued*

19. Measure the plasmid DNA concentration by spectrophotometry ($A_{260} \approx 20.0$ for a 1 mg/ml solution).
20. Digest the plasmid DNA with the appropriate restriction enzyme and verify complete digestion by electrophoresis in an agarose gel (see *Protocol 2*, steps 4–7 and *Protocol 6*).
21. Extract the plasmid DNA by vortexing with saturated phenol:chloroform:isoamyl alcohol.
22. Separate the phases by centrifuging the mixture for 2 min at 10 000 g at room temperature in a microcentrifuge. Remove and retain the upper aqueous phase.
23. Add 0.1 vol. of 3 M sodium acetate, pH 6.5, and 2.5 vol. of absolute ethanol. Mix and leave at $-70\,°C$ for 20 min.
24. Pellet the DNA precipitate in a microcentrifuge (10 min, room temperature).
25. Wash the pellet with absolute ethanol and let it air dry.
26. Resuspend the digested plasmid DNA in H_2O at 1 µg/µl (for a 5 kbp recombinant plasmid) and store it at $-20\,°C$.

[a] When harvesting plasmid DNA from a sealed tube the pressure must first be equilibrated by inserting a hypodermic needle into the top of the tube. The 18 gauge × 1½ inch needle is then inserted (bevelled side up) slightly below the plasmid band and the DNA is collected by aspiration using the syringe.

The disadvantage of the conventional method for preparing template DNA described in *Protocol 1* is the necessity to clone the sequence encoding the desired RNA into the plasmid. Furthermore, the method requires the presence of a restriction site near to the desired termination point for transcription.

One recent breakthrough in nucleic acid technology has been the development of the polymerase chain reaction (PCR) which has now found application in almost every field of research in molecular biology. It is, therefore, not surprising that it has also yielded protocols aimed at providing more versatility in the preparation of template DNA. A procedure based on PCR technology is illustrated in *Figure 1*. It uses two deoxyoligonucleotide primers to amplify a specific region of DNA located between them which is to act as the template for RNA synthesis. Since SP6, T7, and T3 promoters comprise very small sequences of DNA, the first (5′) primer can contain the promoter sequence as well as the sequence homologous to the 5′ end of the transcript. In the example illustrated in *Figure 1*, the first primer contains the T7 promoter sequence as well as 20 nucleotides corresponding to the 5′ end of the transcript. The second (3′) primer is complementary to the 3′ end of the sense strand and thus defines the 3′ end of the transcript. The clear advantages of the procedure are:

1: Synthesis and purification of RNA substrates

(a) It allows for the preparation of large amounts of DNA template from a sequence that does not have to be cloned into a specialized vector with a bacteriophage promoter.

(b) The preparation of template DNA by PCR allows for a total flexibility in the selection of the 5' and 3' ends of the transcript. By judicious choice

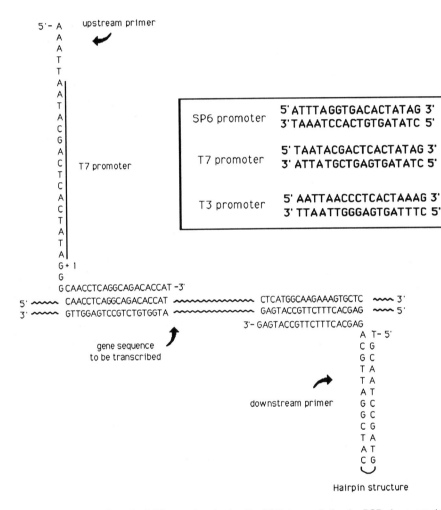

Figure 1. Preparation of a DNA template for *in vitro* RNA transcription by PCR. An example is shown to illustrate the application of this technique. The upstream deoxyoligonucleotide (44 nt) contains the T7 promoter as well as additional sequences complementary to the 5' end of the gene sequence which is to be transcribed. The G at position +1 represents the first nucleotide of the template. The downstream primer (44 nt) contains 20 nt complementary to the 3' end of the gene as well as a sequence capable of forming a hairpin structure designed to protect the transcripts from exonucleolytic degradation. The boxed sequences represent the minimum 18 nt promoter sequences recognized by the three commonly-used bacteriophage RNA polymerases.

of the 5' and 3' oligonucleotide primers for PCR amplification, the 5' and 3' ends of the transcript can be determined with precision.

(c) An advantageous feature of the downstream primer can be the inclusion of a sequence that folds into a hairpin loop, thereby protecting the RNA eventually synthesized from the template from 3' exonuclease attack (see *Figure 1*).

The procedure based on PCR is described in *Protocol 2*. However, the cost of preparing the oligonucleotide primers is high. This may be warranted if the DNA template is needed on a regular basis, but may become prohibitive when transcription reactions are programmed from many different templates. An alternative strategy which produces RNA transcripts with a similar flexibility in the determination of the 3' end involves the use of RNase H after (or during) the transcription reaction (see Section 2.2).

Protocol 2. Preparation of the DNA template by PCR[a]

Equipment and reagents

- 10 × PCR buffer (0.5 M KCl, 0.1 M Tris–HCl, pH 8.0, 15 mM $MgCl_2$)
- dNTP mixture (10 mM each of dATP, dGTP, dCTP, dTTP in 1 mM Tris–HCl, pH 7.5)
- *Taq* DNA polymerase (5 units/μl, Pharmacia or Cetus)
- PCR primers (25 pmol/μl in H_2O)[b]
- DNA sequence encoding the RNA to be transcribed (30 pmol/ml in H_2O); this DNA can be in the form of a purified restriction fragment or be part of a recombinant DNA molecule (e.g. plasmid, cosmid); if the DNA is in the form of a circular plasmid, denature the DNA before use by boiling it for 3 min followed by quick-cooling on ice.
- Temperature cycling apparatus for PCR (Perkin Elmer, MJ Research Inc.)
- Materials for agarose gel electrophoresis (gel box, mould, comb, UV light box)
- 0.5 × TBE buffer (45 mM Tris base, 45 mM boric acid, 1 mM EDTA)
- Electrophoresis dye mixture (50% glycerol, 1 mM EDTA, 0.2% bromophenol blue, 0.2% xylene cyanol FF)

Method

1. Prepare the following mixture (sufficient for 8 reactions):
   - 10 × PCR buffer   16 μl
   - dNTP mixture   3.25 μl
   - *Taq* DNA polymerase (5 units)   1 μl
   - each PCR primer (25 pmol)   1 μl
   - H_2O to 160 μl

2. Distribute the mixture in 20 μl aliquots in 0.5 ml microcentrifuge tubes and add 0.5 μl of the DNA to be amplified to each. Mix and overlay with 4 drops of paraffin oil.

Continued

1: Synthesis and purification of RNA substrates

Protocol 2. *Continued*

3. Amplify the desired DNA sequence using 'touchdown' PCR (5) which consists in performing a first cycle at 94 °C for 30 sec (denaturation), 72 °C for 30 sec (annealing), 72 °C for 1 min (elongation), and then reducing the annealing temperature from 72 °C to 50 °C in steps of 2 °C (use 2 cycles at each temperature). End with 25 cycles at 50 °C. This PCR procedure has the advantage of reducing the effect of spurious priming due to non-specific primer hybridization at low annealing temperatures.
4. Meanwhile prepare a horizontal gel using normal or low melting point agarose (1.0% agarose for an amplified fragment of 0.5–1 kbp in length) in 0.5 × TBE buffer and containing 0.4 µg/ml of ethidium bromide (see *Protocol 6*).
5. Add 2 µl of electrophoresis dye mixture to each PCR reaction and mix.
6. Load the samples onto the agarose gel and electrophorese at 100 volts in 0.5 × TBE buffer until the dyes are well separated.
7. Locate the PCR product by placing the agarose gel on a UV light box. Wear protective eye glasses.
8. Excise the band with a clean scalpel blade. Extract (if using low melting point agarose) or electroelute (if using normal agarose) the DNA using standard procedures (see reference 1).
9. Resuspend the DNA in H_2O at 0.1 µg/µl (for a 500 bp DNA fragment) and store it at −20 °C.

[a] Communicated by I. C. Eperon and based on ref. 4.
[b] One of the most important points to consider when designing a pair of primers is that they should have equivalent melting temperatures (see Chapter 3 for additional hints).

2.2 Transcription using SP6, T3, and T7 RNA polymerases

RNA products of discrete length are usually generated using a 'run-off' transcription reaction on linearized plasmid DNA (see *Protocol 1*) using T7, T3, or SP6 RNA polymerase (1, 4, 5). In this reaction, transcription termination occurs when the RNA polymerase reaches the 3' end of the DNA molecule (see *Figure 2*). However, as mentioned in Section 2.1, it may not always be possible to find a restriction site exactly near the position preferred for termination. Amplification of the DNA template by PCR may provide a solution to this problem (see *Protocol 2*).

For some applications the RNA product may be required in unlabelled form (e.g. for competition experiments). In addition, when problems are encountered with the production of labelled RNA, synthesizing unlabelled RNA provides a fast and inexpensive alternative for testing various transcription conditions. The procedure for synthesizing unlabelled RNA is described in *Protocol 3*. *Figure 3* shows typical experimental results using *Protocol 3*.

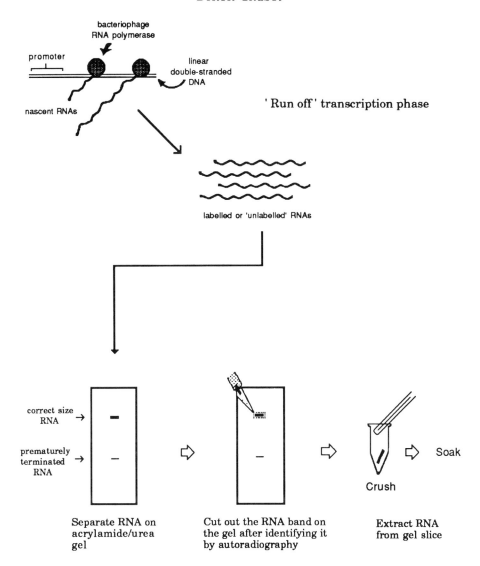

Figure 2. Schematic diagram of the procedure used to prepare and purify *in vitro* RNA transcripts. The DNA template is prepared by digesting recombinant plasmid DNA or by PCR amplification as described in *Protocol 1* and *Protocol 2*, respectively. 'Run-off' transcription using SP6, T7 or T3 RNA polymerase is performed with labelled nucleotides as described in *Protocol 4*. 'Unlabelled' RNA is synthesized with tracer amounts of labelled nucleotide (see Section 2.3.3). Following an optional step of DNase digestion (see *Protocol 6*), the RNA transcripts are then separated by electrophoresis on a denaturing acrylamide gel as described in *Protocol 8*. After visualization of the RNA by autoradiography, the band is cut out and the RNA extracted using the 'crush and soak' procedure described in *Protocol 8*.

1: Synthesis and purification of RNA substrates

Protocol 3. Preparation of unlabelled RNA by SP6, T3, and T7 transcription

Reagents

- 1 M dithiothreitol (DTT)
- 5 × STT transcription buffer mixture:

1 M Tris-HCl, pH 7.5[a]	2 ml
1 M MgCl$_2$[a]	0.3 ml
1 M spermidine[b]	10 µl
H$_2$O[a]	7.69 ml

- rNTPs solution: 5 mM rATP, 5 mM rCTP, 5 mM UTP, and 5 mM rGTP[c]
- Placental RNase inhibitor (e.g. RNAguard™, Pharmacia, approximately 35 units/µl)
- DNA template: either restriction digested plasmid DNA or PCR amplified as described in Protocol 1 or Protocol 2, respectively[d]
- SP6, T3, or T7 RNA polymerase (e.g. USB, Promega, Pharmacia, or Amersham at 10–50 units/µl)

Method

1. Mix 5 µl of 1 M DTT with 95 µl of 5 × STT transcription buffer.
2. Bring the following solutions to **room temperature** and mix them in the order given such that the 5 × STT buffer containing DTT and the DNA are added first and last, respectively[e]

   - 5 × STT transcription buffer containing DTT (from step 1)   5.0 µl
   - rNTPs solution   2.5 µl
   - H$_2$O[a]   13.5 µl
   - placental RNase inhibitor   0.5 µl
   - template DNA[d]   2.5 µl

3. Add 1.0 µl of SP6 or T7 or T3 RNA polymerase and mix gently.
4. Incubate for 1 h at 37 °C. Incubation at 40 °C results in an improved yield with SP6 RNA polymerase (ref. 2).
5. Purify the RNA product as described in Section 2.3.

[a] Filter sterilize these reagents as described in Section 1.1.
[b] Spermidine is hygroscopic and should be dissolved directly in the manufacturer's bottle without weighing.
[c] If the reason for making unlabelled RNA is part of a strategy to solve any problems associated with the synthesis of capped and labelled RNA, the addition of a cap analogue, a low concentration of rGTP (0.1 mM final) and UTP (10 µM final) can be used here to reproduce conditions for making capped and labelled RNA (see Protocol 4).
[d] The concentration of the bacteriophage promoter in the reaction is about 30 nM (2). In a 25 µl transcription reaction, this concentration is attained by adding 2.5 µl of a 5 kbp plasmid DNA at 1 µg/µl (prepared following Protocol 1) or 2.5 µl of a 500 bp PCR-amplified DNA at 0.1 µg/µl (prepared following Protocol 2).
[e] These precautions are necessary to avoid DNA precipitation by spermidine at low temperature.

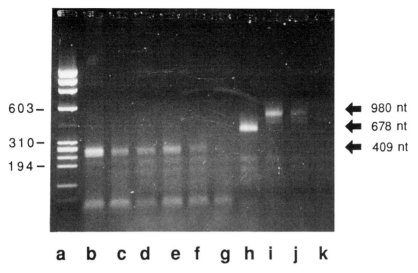

Figure 3. 'Run-off' transcription reactions using SP6 RNA polymerase. Different conditions for SP6 transcription of pSPAd cut either with *Hinc*II (lanes b–g; 409 nt RNA), *Ava*II (lane h; 678 nt) or *Acc*I (lanes i–k; 980 nt) were tested. Transcription reactions were performed exactly as described in *Protocol 3* except that the UTP concentration equalled 20 µM (lanes b, h, and i), 10 µM (lanes c, d, e, and j), 5 µM (lanes f and k) or 1 µm (lane g). The final spermidine concentration was 2 mM throughout, except for the reaction shown in lane c which was performed at 4 mM. The mixing of the ingredients was performed on ice for the reaction shown in lane d. Following transcription, the samples were treated with DNase I (see *Protocol 7*) and the RNA electrophoresed in a horizontal 2% agarose/TBE buffer gel containing ethidium bromide (see *Protocol 6*). ɸX174 DNA digested by *Hae*III was used as DNA molecular weight markers (lane a). The position and size of each RNA transcript are shown arrowed on the right.

Although unlabelled RNA transcripts are needed on occasion, most applications require labelled RNA molecules. Thus, RNA labelled at high specific activity considerably accelerates the analysis of *in vitro* RNA processing events. Since *in vitro* analyses of this kind may involve prolonged incubation of the RNA substrate with cell extracts, the presence of nucleases can lead to non-specific cleavage. Although the presence of RNase inhibitors during the incubation alleviates this problem to some extent, an additional measure of protection against the action of some RNases is to use RNA substrates that are capped at their 5' end. These are produced by including the cap analogue diguanosine triphosphate, G(5')ppp(5')G, in the transcription reaction. At an appropriate concentration relative to rGTP, G(5')ppp(5')G is readily incorporated at the first position (if G is the first nucleotide of the template) by SP6, T7, and T3 RNA polymerases. When a capped RNA is subsequently incubated in a cell extract, the cap structure not only confers protection against 5' exonucleolytic attack but may also contribute to the efficiency and/or the specificity of maturation processes (6–9). As discussed in Section 2.1, the 3' end of the RNA can be protected by using an appropriate template DNA

1: Synthesis and purification of RNA substrates

produced by PCR to generate an RNA hairpin secondary structure that may retard the progress of 3' exonucleases (ref. 4 and see *Figure 1*). The synthesis of capped and labelled RNA is described in *Protocol 4*.

Protocol 4. Preparation of capped and labelled RNA by SP6, T3, and T7 transcription

Reagents

- [α-^{32}P]UTP ($\approx$ 800 Ci/mmol, 20 mCi/ml, $\approx$ 25 µM, Amersham)
- 10 mM G(5')ppp(5')G (in H$_2$O, Pharmacia)[a]
- rXTPs '551' solution (5 mM rATP, 5 mM rCTP, and 1 mM rGTP)
- 0.1 mM UTP
- 5 × STT (containing DTT), placental RNase inhibitor and SP6, T3, or T7 RNA polymerase (see *Protocol 3*)
- DNA template; either restriction digested plasmid DNA or PCR amplified as described in *Protocol 1* or *Protocol 2*, respectively[b]

Method

1. Prepare a working solution of 5 × STT buffer (containing DTT) as described in *Protocol 3*, step 1.
2. Bring the following solutions to **room temperature** and mix them in the order given, such that the 5 × STT transcription buffer and the DNA are added first and last, respectively[c].
   - 5 × STT transcription buffer   4.0 µl
   - rXTPs '551' solution   2.5 µl
   - 0.1 mM UTP   2.5 µl
   - [α-^{32}P]UTP[d]   5.0 µl
   - 10 mM G(5')ppp(5')G   1.0 µl
   - H$_2$O   5.0 µl
   - placental RNase inhibitor   0.5 µl
   - template DNA[b]   2.5 µl
3. Add 1.0 µl of SP6 or T7 or T3 RNA polymerase and mix gently.
4. Incubate for 1 h at 37 °C for T3 and T7 polymerases or at 40 °C for SP6 polymerase.
5. Purify the RNA products as described in Section 2.3.

[a] The dinucleotide m^{7}G(5')ppp(5')G can also be used as a cap analogue. The ratio of cap analogue to rGTP required for optimal cap incorporation is approximately 5 to 1. If the total concentration of rGTP is lowered because labelling is performed with [α-^{32}P]rGTP (see footnote *d*), the amount of cap analogue used should be decreased accordingly.
[b] see *Protocol 3*, footnote *d*.
[c] see *Protocol 3*, footnote *e*.
[d] For making transcripts labelled at higher specific activity, use [α-^{32}P]UTP alone without unlabelled UTP (5 µl of [α-^{32}P]UTP with the specific radioactivity indicated in the *Reagents* list above in a total volume of 25 µl provides a final UTP concentration of 5 µM). If necessary, other labelled nucleotides can be used in combination or instead of [α-^{32}P]UTP. In principle, [α-^{32}P]rGTP is an excellent choice since the rate of SP6 transcription in the presence of low ribonucleotide concentrations is best with this nucleotide (2). Inhibition of transcription sometimes associated with specific batches of [^{32}P]rGTP can be overcome by first lyophilizing the [^{32}P]rGTP followed by resuspension in water. Lyophilization can also be used to increase the concentration of labelled nucleotide in the transcription reaction.

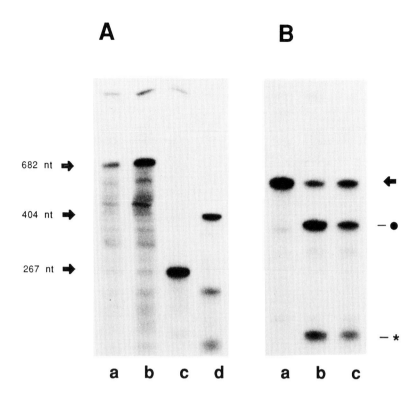

Figure 4. Preparation of labelled RNA by 'run-off' transcription *in vitro*. (A) Transcription by T7 RNA polymerase was carried out from murine *myb* (682 nt, lanes a and b) or *Hox*2.2 (267 nt, lane c; 404 nt, lane d) cDNA fragments following *Protocol 4*, except that the reaction loaded in lane b was performed with [α-^{32}P]UTP alone (final concentration 5 µM). Full-length transcripts are indicated by arrows. (B) Use of RNase H to generate SP6 transcripts from a human β-globin gene. In lane a, transcription was performed as described in *Protocol 4*. In lane b, the transcription reaction contained RNase H, as described in *Protocol 5*. In lane c, RNase H digestion was performed following the transcription reaction by adding 0.5 µl RNase H and 0.5 µl of complementary deoxyoligonucleotide and then incubating for 30 min at 30 °C. In both lane b and lane c, the 507 nt globin RNA (arrow) is cleaved by RNase H to generate two fragments of 326 nt (●) and 161 nt (*).

For most purposes, *Protocol 4* is suitable for generating labelled RNA (see *Figure 4A*). As an alternative to providing the correct 3' termination site by restriction (see *Protocol 1*) or by using a PCR template (see *Protocol 2*), a longer RNA molecule produced by 'run-off' transcription can be cut with RNase H in the presence of a synthetic deoxyoligonucleotide complementary to a specific internal region of the transcript (see *Figure 5*). RNase H will digest the RNA moiety of the RNA:DNA duplex. The use of a transcriptional termination site located well downstream (>50 bp) from the site to be produced by RNase H digestion will allow unambiguous identification of the correct product following gel separation (see *Figure 4B*). Two variations are possible. Either RNase H treatment occurs simultaneously with transcription (see *Protocol 5*) or the treatment with the oligonucleotide and RNase H occurs after *in vitro* transcription. This latter strategy may be necessary if for instance one needs to control non-specific hybridization of the oligonucleotide at other regions of the RNA. This may be done by raising the stringency or temperature of hybridization prior to RNase H digestion or by digesting for a shorter period of time.

Protocol 5. Generation of RNA substrates with specific 3' termini using RNase H digestion

Reagents

- Deoxyribonucleotide complementary to the desired region of the RNA product
- RNase H (Pharmacia, 0.8 units/µl)
- Transcription reaction mixture (25 µl) ready for incubation (see *Protocol 4*, steps 1–4)

Method

1. As a control for full-length product, transfer a small aliquot of transcription reaction mixture (2.5 µl) into a separate microcentrifuge tube.
2. To the rest of the transcription reaction mixture, add
   - 100 pmol of a complementary deoxyoligonucleotide (e.g. 0.5 µg of a 15 nt oligomer)
   - 1 µl of RNase H
3. Incubate at 37 °C for 1 h.
4. Purify the RNA product as described in Section 2.3.

2.3 Purification of *in vitro* RNA transcripts

2.3.1 Examination of unlabelled RNA products made *in vitro*

The goal of transcription reactions performed with unlabelled nucleotides may be to synthesize unlabelled RNA to be used as a competitor in RNA processing experiments *in vitro* or to test various conditions for solving problems associated with the transcription of certain templates. In both cases, electrophoresis of unlabelled RNA in non-denaturing agarose gels provides a rapid procedure for evaluating the efficiency of the transcription reaction and for verifying the size of the RNA produced. Only a fraction of the 25 µl transcription reaction (e.g. 5 µl) obtained in *Protocol 3* need be loaded on an agarose gel to verify the homogeneity of the transcripts, while the rest of the transcription reaction can then be purified following the protocols described in this section. The separation and observation of unlabelled RNA produced by *in vitro* transcription is described in *Protocol 6*.

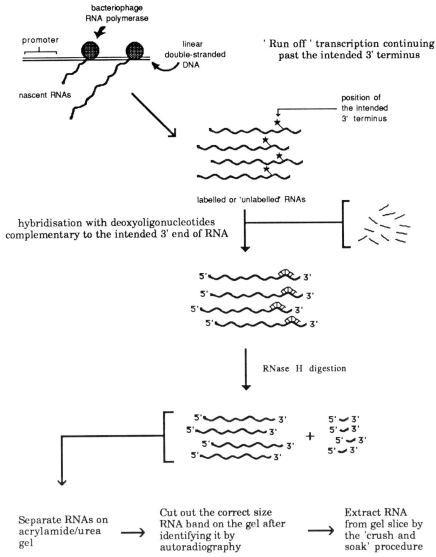

Figure 5. Schematic diagram of the procedure used to generate and purify RNA transcripts with specific 3' termini using RNase H digestion. The DNA template is prepared by digesting recombinant plasmid DNA as described in *Protocol 1*. 'Run-off' transcription using SP6, T7, or T3 RNA polymerase is performed as described in *Protocol 4* and Section 2.3.3 and continues past (>50 nt) the intended 3' terminus. 'Unlabelled' RNA is synthesized with tracer amounts of labelled nucleotide (see Section 2.3.3). Deoxyoligonucleotides complementary to the position intended as the 3' terminus are then added to the transcription mixture as well as RNase H which will digest the RNA moiety of the RNA:DNA hybrid. RNase H treatment can also be accomplished simultaneously with transcription as described in *Protocol 5*. The RNA molecules generated by transcription and RNase H digestion are then separated by electrophoresis on an acrylamide/urea gel as described in *Protocol 8* and shown in *Figure 4B*. After visualization of the RNA, the band is cut out and the RNA extracted using the 'crush and soak' procedure described in *Protocol 8*.

1: Synthesis and purification of RNA substrates

Procotol 6. Examination of unlabelled RNA by agarose gel electrophoresis

Equipment and reagents

- 0.5 × TBE buffer and electrophoresis dye mixture[a] as described in *Protocol 2*
- Horizontal agarose gel electrophoresis apparatus
- 1.5% or 2% agarose gel containing 0.5 × TBE buffer and 0.4 µg/ml ethidium bromide
- RNA sample for analysis (i.e. transcription reaction mixture, 25 µl, from *Protocol 3*)

Method

1. Add 2.5 µl of electrophoresis dye mixture to the 25 µl transcription reaction mixture[b] and mix.
2. Load the sample directly on to the horizontal agarose gel together with appropriate DNA size markers and electrophorese (15 volts/cm) in 0.5 × TBE buffer until good separation of the dye, or the DNA markers, has occurred. Since the gel contains ethidium bromide, the extent of DNA separation can be verified at anytime using an UV light source. By comparison to double-stranded DNA markers, a typical RNA without extensive secondary structure runs at a position approximately equivalent to a DNA half its size in base pairs (see *Figure 3*).
3. Locate the RNA product by UV transillumination (see *Protocol 2*, step 7).

[a] Filter sterilized as described in Section 1.1.
[b] If the DNA used in the transcription reaction (digested plasmid or PCR amplified) contains fragments that may interfere with the detection of the unlabelled RNA transcript, digest the transcription reaction mixture with DNase I according to *Protocol 7*, step 1.

2.3.2 Purification of homogeneous RNA products

Transcripts prepared by SP6, T7, or T3 transcription are often very homogeneous with little or no shorter products. When this is the case, digestion with DNase I followed by extraction with organic solvents and ethanol precipitation may yield RNA that is sufficiently pure for most applications. Thus, for RNA produced by the procedures in Section 2.2, the purification described in *Protocol 7* may suffice.

Protocol 7. Purification of *in vitro* RNA substrates by DNase treatment and ethanol precipitation

Reagents

- Transcription reaction mixture (25 µl) containing RNA transcripts (from *Protocols 3* or *4*)
- DNase I (RNase-free, molecular biology grade, 5 units/µl, Pharmacia)

Continued

Protocol 7. *Continued*

- 0.3 M sodium acetate, 0.2% SDS (filter sterilized see Section 1.1)
- Phenol:chloroform:isoamyl alcohol (25:24:1) saturated in 10 mM Tris–HCl, pH 7.5, 1 mM EDTA
- Glycogen (molecular biology grade; Calbiochem, 25 µg/µl in H_2O)

Method

1. Add 1 µl (5 units) of DNase I to the 25 µl transcription reaction mixture and incubate for 15 min at 37 °C.
2. Add 300 µl of 0.3 M sodium acetate, 0.2% SDS, and 300 µl of saturated phenol:chloroform:isoamyl alcohol. Extract the RNA by vortexing.
3. Separate the phases in a microcentrifuge. Remove and retain the upper aqueous phase.
4. To help recover small quantities of RNA, add 1 µl of glycogen solution as a carrier[a].
5. Add 2.5 vol. of absolute ethanol. Mix and leave at −70 °C for 20 min.
6. Pellet the RNA in a microcentrifuge (10 min, room temperature).
7. Wash the pellet with absolute ethanol and air dry it.
8. Resuspend the dried RNA in 20 µl sterile H_2O and store it at −70 °C.

[a] The use of glycogen does not appear to interfere with any subsequent processing reactions in which the RNA may be used as a substrate.

2.3.3 Purification of *in vitro* RNA substrates by polyacrylamide/urea gel electrophoresis

In contrast to the ideal situation, *in vitro* transcription by phage RNA polymerases may yield, in addition to the expected transcript, substantial levels of shorter RNA products (see *Figure 4A*) necessitating subsequent purification of the correct product. Premature polymerase termination can arise from a variety of reasons.

(a) One possibility is that the DNA template contains sequences that resemble termination signals normally used by the bacteriophage RNA polymerase. Obviously, the longer the expected transcript, the greater the chances that such sequences will be encountered.

(b) Premature termination also occurs more frequently with radiolabelled RNA especially when the ribonucleotide concentration used is lowered (< 10 µM ribonucleotide) to favour the synthesis of high specific activity transcripts (see *Protocol 4*, footnote *d*).

(c) Variations in the purity of the template DNA can also influence the rate at which RNA polymerases prematurely terminate.

1: Synthesis and purification of RNA substrates

(d) SP6 RNA polymerase has been documented to 'stutter' in regions containing extensive stretches of uridines, resulting in RNA molecules of heterogeneous size.

In most cases these problems can be overcome after transcription has finished by purifying the correct sized RNA product on a gel. *Radiolabelled* RNA from 100–1000 nt in size is routinely purified by polyacrylamide gel electrophoresis on a 4% polyacrylamide/7.0 M urea gel as described in *Protocol 8*.

Protocol 8. Isolation of *in vitro* RNA substrate by polyacrylamide/urea gel electrophoresis

Equipment and reagents

- 40% acrylamide stock solution (38% acrylamide, 2% methylene bis-acrylamide in H_2O; store it in the dark at 4 °C)
- Urea (reagent grade, there is no need to use RNase-free)
- 10 × TBE buffer (0.9 M Tris base, 0.9 M boric acid, 20 mM EDTA)
- 10% ammonium persulphate (freshly made)
- TEMED
- Electrophoresis formamide–dye mixture (1 mM EDTA, 0.1% bromophenol blue, 0.1% xylene cyanol FF in deionized formamide)
- Vertical polyacrylamide gel electrophoresis equipment
- RNA transcription reaction mixture (25 µl) from *Protocols 3, 4,* or *5*
- A sterilized, siliconized glass rod. Immediately before use, rinse a glass rod with ethanol and flame it, then treat it with a dimethyldichlorosilane solution
- 0.3 M sodium acetate, pH 6.0, 0.2% SDS
- Rotatory shaker (e.g. LabQuake™, Becton Dickinson)
- Saturated phenol:chloroform:isoamyl alcohol (25:24:1) as described in *Protocol 1*

Method

1. For a 50 ml denaturing 4% polyacrylamide gel use:
   - 40% acrylamide   5 ml
   - urea   21 g
   - 10 × TBE buffer   5 ml
   - H_2O to 50 ml
2. Dissolve the urea and then filter the mixture through a Nalgene filter unit (0.2 µm pore size (see Section 1.1).
3. Add 0.5 ml of 10% ammonium persulphate and 30 µl TEMED. Mix well.
4. Quickly pour the mixture between the glass plates (use 1.5 mm thickness spacers) used to form the gel mould and allow the gel to polymerize.
5. Mount the gel sandwich on to the electrophoresis apparatus.
6. Add 12.5 µl of electrophoresis dye mixture to the RNA transcription reaction and mix.
7. Heat the solution at 100 °C for 90 sec.

Continued

Protocol 8. *Continued*

8. Immediately before loading the samples, rinse the sample wells of the gel with 1 × TBE buffer to remove any urea which may have diffused into them.
9. Run the gel until the dyes are well separated[a].
10. Remove one glass plate leaving the gel on the other plate. Place a sheet of plastic cling film over the gel.
11. Mark the gel with radioactive or phosphorescent ink in an asymmetrical fashion to aid orientation after autoradiography. This is best done by applying self-adhesive labels spotted with dried ink on the cling film.
12. Expose the gel to a Kodak XRP film for 15 sec and develop the film.
13. Cut out the appropriate region of the gel containing the desired RNA band by placing the gel on top of the autoradiogram using the ink markers to align its position. Cut out the gel slice corresponding to the desired band with a clean and sharp scalpel blade, carefully disposing of the small piece of cling film, and transfer the gel slice into a sterile microcentrifuge tube.
14. Crush the acrylamide until it is in fine pieces, using the sterile siliconized glass rod.
15. Add 0.4 ml of 0.3 M sodium acetate, 0.2% SDS. Place the tube on a rotatory shaker and gently shake for at least 3 h at room temperature (a recovery of greater than 80% is typically achieved after 3 h for a fragment ~500 nt in length).
16. Pellet the acrylamide for 2 min in a microcentrifuge.
17. Remove the aqueous phase to a clean microcentrifuge tube.
18. Add 400 µl of saturated phenol : chloroform : isoamyl alcohol. Vortex to mix the phases and then centrifuge for 2 min in a microcentrifuge.
19. Precipitate the aqueous phase with 2.5 vol. of ethanol as described in *Protocol 7*, steps 5–7[b].
20. Redissolve the RNA in 20 µl sterile H_2O and store it at −70 °C.

[a] On a 4% acrylamide/urea gel, bromophenol blue and xylene cyanol migrate at positions approximately equivalent to RNAs of 50 nt and 150 nt, respectively. For most purposes, migration can be stopped before the unincorporated radioactive nucleotide runs out of the gel, allowing safer disposal of radioactive materials.
[b] There is no need to add carrier RNA or glycogen to help precipitation of the RNA sample. Following precipitation, the RNA pellet is readily visible; possibly due to coprecipitation of salt. The coprecipitated salt does not seem to exert an inhibitory effect in subsequent *in vitro* assays using the RNA substrate.

If the transcripts produced are homogeneous, the purification of *unlabelled* RNA is accomplished by using *Protocol 7*. However, if unlabelled transcripts of different sizes are produced (as visualized on an agarose gel using *Protocol 6*), the recommended purification strategy is to add a very small amount of labelled nucleotide as a tracer to the transcription reaction described in *Protocol 3* without interfering with the original reason for using unlabelled RNA. To do this, add 0.5 µl of [α-^{32}P]UTP (see *Protocol 4*) to the transcription

1: Synthesis and purification of RNA substrates

Table 1. Quantification of the amount of *in vitro* RNA transcripts purified

A. *Amount of ^{32}P-nucleotide in purified RNA*

1. Count an aliquot of the final ^{32}P-labelled RNA sample in a scintillation counter. For convenience, use direct Cerenkov counting. For example, if 1/50th of the total RNA sample contains 100 000 c.p.m., this corresponds to approximately 200 000 d.p.m. (assuming a typical efficiency of Cerenkov counting of 50%) or 1×10^7 d.p.m. for the total RNA sample.
2. Since in *Protocol 4* 5 μl of [α-^{32}P]UTP was used corresponding to 100 μCi (2.2×10^8 d.p.m.), the percentage of [^{32}P]UTP present in the purified labelled RNA is:

$$\frac{1 \times 10^7 \text{ d.p.m.} \times 100}{2.2 \times 10^8 \text{ d.p.m.}} = 4.5\%. \tag{1}$$

B. *Amount of RNA synthesized*

1. If for example, 4.5% of the *labelled* UTP has been incorporated in the purified RNA, it can be assumed that 4.5% of the *total* UTP has also been incorporated. Furthermore, if the transcription reaction in *Protocol 4* contained 2.5×10^{-10} mol of unlabelled UTP (i.e. 2.5 μl of 0.1 mM), and approximately 1.25×10^{-10} mol of labelled UTP (i.e. 5 μl of 25 μM [^{32}P]UTP), the total amount of UTP present was 3.75×10^{-10} mol.
2. In this example, 4.5% of 3.75×10^{-10} mol of UTP total is present in the purified ^{32}P-labelled RNA. This corresponds to 1.7×10^{-11} mol (17 pmol) of UTP.
3. For a transcript of 500 nt, where on average one base in four will be a uracil (i.e. 125 nt), the molar amount of RNA transcript synthesized corresponds to:

$$\frac{1.7 \times 10^{-11} \text{ mol}}{125} = 1.35 \times 10^{-13} \text{ mol}. \tag{2}$$

Hence in this example, a total of 135 fmol of RNA has been purified (~0.022 μg). The specific activity is about 5.6×10^8 d.p.m./μg.

---

reaction specified in step 2 of *Protocol 3*. The specific activity of the labelled RNA produced by this procedure will be 500-fold less than the labelled RNA synthesized following *Protocol 4*. Transcript purification can then be performed as described in *Protocol 8* except that in step 12 the gel is exposed for 2 h.

The separation of RNA molecules longer than 1 kb is problematic if transcription yields heterogeneous products. For such large transcripts, elution using the 'crush and soak' procedure described in *Protocol 8* is not recommended because recovery is not efficient. An alternative procedure involves the rapid purification of the RNA using commercially available purification kits such as RNaid™ (Bio 101); follow the instructions provided by the manufacturer. Alternatively, one can attempt to reduce premature termination by increasing the concentration of the limiting nucleotide and so avoid the need to purify the correct size RNA. However, the disadvantage of this strategy is that it reduces the specific activity of the transcript.

2.3.4 Quantification of RNA substrates following purification

On many occasions, it is important to know with some accuracy the amounts of RNA used in processing reactions (e.g. in competition experiments). The final molar or microgram amount of *labelled* RNA can be calculated as described in *Table 1*. The same procedure is applied to calculate the amount of '*unlabelled*' RNA synthesized in the presence of tracer amounts of labelled nucleotide.

2.3.5 Problem solving

When an *in vitro* transcription reaction does not work or works inefficiently, first check the template DNA. To check for the presence of an inhibitor of transcription, mix the DNA template with another template that has previously worked successfully. Keep the total DNA concentration lower than 2.5 µg per 25 µl of transcription reaction. (*Protocols 3, 4,* and *5*).

The problem may also be an excessive concentration of spermidine in the $5 \times$ STT (see *Protocol 3*). Try transcription reactions with a decreased spermidine concentration (1 mM or 0.25 mM).

3. Purification of RNA from cells

The analysis of maturation processes often requires the purification of RNA from cells and its analysis using Northern blots and various mapping experiments using reverse transcriptase or nucleases (see Chapter 3). More recently, mapping RNA by making use of the PCR procedure following a first step of reverse transcription has become popular (see Chapters 2 and 3). Paramount to the success of all these techniques is the ability to obtain reproducibly undegraded RNA. Use the precautions described in Section 1.1 to protect the RNA from RNases during its isolation.

A variety of methods are available for RNA purification from cells and tissues. Techniques that have improved the rapidity of the extraction process have understandably become preferred. This section presents three of the most rapid, convenient, and reproducible methods for isolating RNA from mammalian cells.

3.1 Purification of total RNA

The first protocol presented (*Protocol 9*) is a modification of the procedure of Auffray and Rougeon (10). This extraction procedure has been used successfully to isolate total RNA from tissues containing high level of RNases (e.g. pancreas) and from tissues highly contaminated with bacteria (e.g. intestine).

1: Synthesis and purification of RNA substrates

It provides a certain way of obtaining intact total RNA from both tissues and cells grown in tissue culture. The cells are disrupted in a solution containing 6 M urea and 3 M LiCl which rapidly inactivates cellular RNases. A sonication step may be included at this stage to break DNA (chromosomal, or plasmid DNA if the cells have been transfected) into smaller pieces and so assure a greater solubility of DNA. Alternatively, the RNA sample may be treated with DNase I. The RNA is then deproteinized and recovered by ethanol precipitation. The procedure described in *Protocol 9* is adapted for extraction of total RNA from multiple samples of tissue culture cells grown in separate Petri dishes (following transfection with different DNA samples, for example).

Protocol 9. LiCl/urea extraction method for total RNA

Reagents

- 6.0 M LiCl/3 M urea solution; prepare this fresh (less than 12 h before use); for 25 ml, mix 3.18 g of LiCl with 9 g of urea (RNase-free urea is not required) and add H_2O to 25 ml; dissolve the urea and then filter the solution through a sterile cellulose nitrate membrane (Nalgene filter unit 120-0020; see Section 1.1); keep on ice
- PBS (0.13 M NaCl, 20 mM potassium phosphate, pH 7.4); keep on ice
- 10 mM Tris-HCl, pH 7.6, 0.5% SDS
- Phenol:chloroform:isoamyl alcohol (25:24:1) saturated with 10 mM Tris-HCl, pH 7.5, 1 mM EDTA

Method

1. Aspirate the medium from each Petri dish of cells.
2. Rinse the cells with cold PBS (10 ml per 100 mm Petri dish).
3. Add 1 ml PBS per dish and harvest the cells using a tissue culture scraper (policeman).
4. Transfer the cells to a microcentrifuge tube and centrifuge for 15 sec.
5. Remove the PBS by aspiration.
6. Resuspend the cell pellet in 1 ml of cold LiCl/urea solution.
7. At this point an optional sonication step may be included. Sonicate the cells 3 times for 40 sec, each time using the maximum frequency setting. Keep the tube in ice during sonication and allow a 20 sec break between each period of sonication.
8. Leave the cell suspension on ice overnight. Shorter incubations do not give efficient RNA recovery.
9. Centrifuge 15 min at 4 °C in a microcentrifuge (12 000 r.p.m., 10 000 g).
10. Carefully remove all the supernatant and resuspend the pellet in 0.4 ml of 10 mM Tris-HCl, pH 7.6, 0.5% SDS.
11. Vortex with an equal volume of buffer-saturated phenol:chloroform:isoamyl alcohol.
12. Separate the phases by centrifugation in a microcentrifuge. Remove and retain the upper aqueous phase.

Continued

Protocol 9. *Continued*

13. Repeat steps 10–12 until the interface is completely clear (two extractions should be enough).
14. Precipitate the RNA with ethanol as described in *Protocol 7*, steps 5–7.
15. Redissolve the RNA in 15 µl of sterile H_2O (approximately 10 µg/µl) and store it at $-70\,°C$.

Other popular methods for isolating total RNA involve extraction with strong denaturants such as guanidine–HCl and guanidinium thiocyanate. A modification of the procedure developed by Chomczynski and Sacchi (11) described in *Protocol 10* uses guanidinium thiocyanate followed by phenol: chloroform:isoamyl alcohol extraction, and shares with *Protocol 9* the advantage of being convenient for the simultaneous processing of several samples. Because this technique does not require the ultracentrifugation step in CsCl solutions usually associated with guanidinium thiocyanate extraction procedures, it allows the isolation of pure total RNA in less than 1 h.

Protocol 10. Guanidinium thiocyanate/phenol/chloroform extraction method for total RNA

Reagents

- GT solution[a] (4 M guanidinium thiocyanate (Fluka), 25 mM sodium citrate, pH 7.0, 0.5% sarcosyl (*N*-lauroyl sarcosine, from a 10% stock solution kept at 65 °C), 0.1 mM 2-mercaptoethanol)
- 3.0 M sodium acetate, pH 4.0
- Phenol saturated with H_2O
- Chloroform:isoamyl alcohol (49:1)
- PBS (0.13 M NaCl, 20 mM potassium phosphate, pH 7.4)

Method

The following procedure is suitable for one Petri dish of cells ($\approx 5 \times 10^6$ cells).

1. Wash the cells with PBS and recover them as described in *Protocol 9*, steps 1–5.
2. Resuspend and lyse the cells by vortexing in 0.4 ml of GT solution.
3. Add 27 µl of 3 M sodium acetate, pH 4.0, and vortex.
4. Add 0.4 ml of phenol (saturated with water) and vortex.
5. Add 80 µl of chloroform:isoamyl alcohol (49:1). Vortex for 10 sec and cool the mixture on ice for 5 min.
6. Centrifuge for 5 min in a microcentrifuge at maximum speed at 4 °C. Following centrifugation, DNA and proteins are distributed at the interphase and in the organic phase while RNA remains in the aqueous phase.
7. Slowly and carefully collect the upper aqueous phase and add an equal volume of isopropanol[b].

Continued

Protocol 10. Continued

8. Incubate for a minimum of 30 min at $-20\ °C$ and then centrifuge for 10 min in a microcentrifuge.
9. Carefully rinse the pellet twice with 75% ethanol and once with absolute ethanol. If, while rinsing, the RNA pellet detaches from the inside of the microcentrifuge tube, centrifuge it briefly (1 min) in a microcentrifuge.
10. Air dry the RNA pellet and dissolve it in 15 µl of sterile H_2O. Store the RNA solution at $-70\ °C$.

[a] **GT solution** can be kept for several months at room temperature but store it in a safe place because guanidinium thiocyanate is highly toxic.
[b] Ethanol is not recommended because it promotes salt precipitation.

3.2 Purification of cytoplasmic RNA

It may become important in the course of investigations on RNA maturation to isolate cytoplasmic rather than total RNA. Some procedures have been described that allow cytoplasmic and nuclear RNA to be obtained from the same cellular preparation (1, 12). *Protocol 11* allows for the rapid purification of intact cytoplasmic RNA using urea to prevent ribonucleolytic activity.

Protocol 11. The NP40/urea extraction procedure for cytoplasmic RNA[a]

Reagents

- Solution A (0.1 M NaCl, 10 mM Tris–HCl, pH 7.9, 0.65% Nonidet P-40 (NP-40)[b]
- Solution B (7.0 M urea[c], 0.35 M NaCl, 10 mM Tris–HCl, pH 7.4, 10 mM EDTA, 1% SDS)[b]
- PBS: 0.13 M NaCl, 20 mM potassium phosphate, pH 7.4
- Phenol:chloroform:isoamyl alcohol (25:24:1) saturated with 10 mM Tris–HCl, pH 7.5, 1 mM EDTA

Method

1. Wash the cells with PBS and recover them as described in *Protocol 9*, steps 1–5.
2. Lyse the cells by thoroughly resuspending (do not vortex) the cell pellet in at least 4 vol. of solution A.
3. Centrifuge the lysate in a microcentrifuge for 30 sec. The pellet contains nuclei.
4. Transfer the supernatant to a microcentrifuge tube that contains an equal volume of solution B.
5. Add an equal volume of saturated phenol:chloroform:isoamyl alcohol, extract and recover the RNA as described in *Protocol 9*, steps 11–15.

[a] Adapted from ref. 13.
[b] Filtered through a cellulose nitrate membrane before use (see Section 1.1).
[c] The use of RNase-free urea is not necessary.

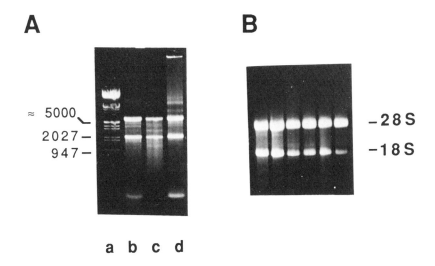

Figure 6. (A) Extraction of RNA from cells. RNA was purified from one Petri dish (100 mm) of HeLa cells (about 5×10^6 cells) using the NP-40/urea method for cytoplasmic RNA (lane b, see *Protocol 11*), or the LiCl/urea (lane c, see *Protocol 9*), or the guanidinium thiocyanate (lane d, see *Protocol 10*) extraction procedures for total RNA. In each case 10% of the RNA extracted was loaded on to a 0.8% agarose gel containing $0.5 \times$ TBE buffer and ethidium bromide (see *Protocol 6*). Lane a contains DNA molecular weight markers (sizes in bp) generated by *Hin*dIII and *Eco*RI digestion of phage lambda DNA. (B) RNA was extracted from various mouse erythroleukaemic cell lines using the LiCl/urea procedure (see *Protocol 9*, sonication step included). The positions of the 28S and 18S ribosomal RNAs are indicated.

As shown in *Figure 6*, the quality and yield of the RNA obtained using the LiCl/urea, the guanidinium thiocyanate, or NP-40/urea extraction procedures (*Protocols 9–11*) can be quickly verified by running a small horizontal 0.8% agarose gel in $0.5 \times$ TBE buffer as described in *Protocol 6*. Under these conditions, the large ribosomal RNAs should not only constitute the major bands (28S and 18S, see *Figure 6*), but also be recovered in approximately equivalent amounts and show little smearing. As a guide, all these methods can be expected to yield approximately 30 µg of total RNA from 10^6 transformed HeLa cells.

3.3 Purification of poly(A)⁺ RNA

The isolation of poly(A)⁺ RNA is often required when the level of a particular mRNA is too low to be detected in a total RNA sample. Several procedures and commercial kits are available to enrich efficiently for poly(A)⁺ RNA. The purification procedure described in *Protocol 12* relies on affinity chromatography of poly(A)⁺ messenger RNA using oligo(dT)–cellulose. The procedure is quick, reproducible, and provides a cheap alternative compared to relatively expensive kits sold by biotechnology companies. Typically, 3% of

1: Synthesis and purification of RNA substrates

the total RNA is recovered in the poly(A)$^+$ fraction following a single purification round.

An increasingly popular and convenient way to purify poly(A)$^+$ RNA from several samples is through the use of an affinity paper commercially known as Hybond™-mAP (Amersham). In this case follow the instructions provided by the manufacturer.

Protocol 12. Purification of poly(A)$^+$ RNA by oligo(dT)–cellulose chromatography

Reagents

- 2× and 1× loading buffer (1× loading buffer is 20 mM Tris-HCl, pH 7.6, 0.5 M NaCl, 1 mM EDTA, 0.1% SDS)
- 0.1 M NaOH, 5 mM EDTA
- Elution buffer (10 mM Tris-HCl, pH 7.6, 1 mM EDTA, 0.05% SDS)
- Total or cytoplasmic RNA in sterile H$_2$O
- ($\leqslant$10 μg/μl) prepared as described in *Protocols 9–11*.
- Oligo(dT)–cellulose (Pharmacia, Boehringer Mannheim)
- Glycogen (molecular biology grade; 25 μg/μl; Calbiochem)
- 3.0 M sodium acetate, pH 4.0

Method

Fractionate amounts of RNA $\geqslant$ 500 μg on oligo(dT)–cellulose packed in sterile disposable columns (Poly-Prep® chromatography columns, Bio-Rad). For smaller quantities, carry out the procedure in sterile microcentrifuge tubes; this allows the simultaneous processing of several RNA samples.

A. *Preparation of the oligo(dT)–cellulose*

Approximately 250 μl of packed oligo(dT)–cellulose is needed for each milligram of total RNA to be processed. The following steps can be performed directly in the column or in microcentrifuge tubes. The whole procedure is carried out at room temperature.

1. Resuspend the oligo(dT)–cellulose in 5 vol. of 1× loading buffer.
2. Wash the oligo(dT)–cellulose with 0.1 M NaOH, 5 mM EDTA followed by several washes with sterile H$_2$O until the pH of the washings is <pH 8.0.
3. Wash the oligo(dT)–cellulose twice with 1× loading buffer.

B. *Purification of poly(A)$^+$ RNA*

1. Heat the total or cytoplasmic RNA sample at 65 °C for 3 min and then rapidly chill it on ice.
2. Add one vol. of 2× loading buffer.
3. Follow *i* or *ii*.

i. *Column procedure*

4. Slowly load the RNA sample on to the column. Since the volume of the RNA sample will be smaller or equal to that of the oligo(dT)–cellulose, stop the flow of the column as soon as the RNA sample has completely entered the column.

Continued

Protocol 12. *Continued*

5. Leave the column to stand for 1 min.
6. Wash the column four times with 1 ml of 1 × loading buffer each time.
7. Elute the poly(A)$^+$ RNA with 4 × 1 ml lots of elution buffer.
8. Add 3 M sodium acetate, pH 4.0, to a final concentration of 0.3 M, followed by 200 µg of glycogen.
9. Add 2.5 vol. of absolute ethanol. Mix and leave at −70 °C for 20 min.
10. Recover the RNA precipitate by centrifugation in a microcentrifuge (10 min, room temperature).
11. Wash the pellet with absolute ethanol and let it air dry. Resuspend the RNA in 20 µl of sterile H$_2$O and store it at −70 °C.

or

ii. *Microcentrifuge tube procedure*

4. Add the RNA sample (from step 2) to the appropriate amount of prepared oligo(dT)-cellulose in a microcentrifuge tube.
5. Mix gently (on a rotator) at room temperature for 5 min.
6. Centrifuge for 4 sec in a microcentrifuge. Discard the supernatant.
7. Wash the oligo(dT)-cellulose four times by resuspension and centrifugation with 1 ml of 1 × loading buffer each time.
8. Elute the poly(A)$^+$ RNA twice with elution buffer. For each elution step use a volume of elution buffer equivalent to five times the volume of the packed oligo(dT)-cellulose.
9. Add 3 M sodium acetate, pH 4.0, to a final concentration of 0.3 M followed by 30 µg of glycogen per sample.
10. Add 2.5 vol. of absolute ethanol, vortex to mix and leave the mixture at −70 °C for 20 min. Recover and store the RNA as described in steps 10–11 of the column procedure described above.

4. Purification of snRNAs

Small nuclear RNAs (snRNAs) are components of small nuclear ribonucleoprotein particles (snRNPs) and are thus important protagonists in RNA maturation processes. Assessing the role of individual snRNAs in processing events has been accomplished by a variety of procedures that have included immunodepletion, snRNA degradation, and snRNP inactivation and depletion using specific oligonucleotides (see Chapter 4). The preparation of the major snRNA species for the purpose of obtaining sequence information has traditionally involved immunoprecipitation with anti-snRNP antibodies, then resolution of the snRNAs by acrylamide/urea gel electrophoresis followed by the standard 'crush and soak' extraction procedure (see *Protocol 8*). Minor snRNA species typically require separation by two-dimensional gel fractionation (14, 15). The immunoprecipitation step can sometimes be advantageously replaced by fractionation of the extract on a CsCl gradient using the protocol described by Lelay-Taha et al. (16). This procedure yields snRNA fractions devoid of tRNAs, 7S, 5S, and large rRNAs, and even accomplishes some separation of the snRNP populations.

It has not yet been possible to demonstrate that snRNPs assembled *in vitro* are functional, although snRNAs purified from denaturing acrylamide gels are properly assembled after injection into *Xenopus* oocytes and are then functional (17). Therefore, it has been necessary to purify snRNP particles from cells in order to address their role in specific RNA processing events. The isolation of snRNPs by biochemical fractionation and affinity chromatography is discussed in Chapter 4 and immunological approaches to snRNP interactions are discussed in Chapter 5.

5. Purification of other RNAs

The bulk of the RNA removed during the purification of poly(A)$^+$ RNA by affinity chromatography comprises the ribosomal RNAs which make up more than 85% of total cellular RNA. Chapter 5 in Volume II discusses various aspects related to the purification of these ribosomal RNAs and their post-transcriptional processing. Investigators interested in the isolation and purification of tRNA and yeast RNA should consult Chapter 6, this volume and Chapter 6, Volume II.

Acknowledgements

The author is grateful to Ian Eperon and Louise Bouchard who have freely communicated helpful information about the protocols and thanks Ian Eperon and members of the author's laboratory for their helpful comments.

References

1. Sambrook, J., Fritsch, E. F., and Maniatis, T. (ed.) (1989). *Molecular cloning: a laboratory manual*. Cold Spring Harbor Laboratory Press, Cold Spring Harbor, NY.
2. Melton, D. A., Krieg, P. A., Rebagliati, M. R., Maniatis, T., Zinn, K., and Green, M. R. (1984). *Nucl. Acids Res.*, **12**, 7035.
3. Davenloo, P., Rosenberg, A. H., Dunn, J. J., and Studier, F. W. (1984). *Proc. Natl Acad. Sci. USA*, **81**, 2035.
4. Cunningham, S. A., Else, A. J., Potter, B. V. L., and Eperon, I. C. (1991). *J. Mol. Biol.*, **217**, 265.
5. Don, R. H., Cox, P. T., Wainwright, B. J., Baker, K., and Mattick, J. S. (1991). *Nucl. Acids Res.*, **19**, 4008.
6. Konarska, M. M., Padgett, R. A., and Sharp, P. A. (1984). *Cell*, **38**, 731.
7. Krainer, A. R., Maniatis, T., Ruskin, B., and Green, M. R. (1984). *Cell*, **36**, 993.
8. Edery, I. and Sonenberg, N. (1985). *Proc. Natl Acad. Sci. USA*, **82**, 7590.
9. Inoue, K., Ohno, M., Sakamoto, H., and Shimura, Y. (1989). *Genes Dev.*, **3**, 1472.
10. Auffrey, C. and Rougeon, F. (1980). *Eur. J. Biochem.*, **107**, 303.
11. Chomczynski, P. and Sacchi, N. (1987). *Anal. Biochem.*, **162**, 156.

12. Chang, D. D. and Sharp, P. A. (1989). *Cell*, **59**, 789.
13. Berk, A. J. and Sharp, P. A. (1977). *Cell*, **12**, 721.
14. Tollervey, D. (1987). *J. Mol. Biol.*, **196**, 355.
15. Montzka, K. A. and Steitz, J. A. (1988). *Proc. Natl Acad. Sci. USA*, **85**, 8885.
16. Lelay-Taha, M.-N., Réveillaud, I., Sri-Widada, J., Brunel, C., and Jeanteur, Ph. (1986). *J. Mol. Biol.*, **189**, 519.
17. Pan, Z.-Q. and Prives, C. (1988). *Science*, **241**, 1328.

2

Characterization of RNA

PAULA J. GRABOWSKI

1. Introduction

Characterization of RNA is necessary to identify the transcripts of newly-described genes, to establish the pattern of splicing and 3′ end formation, to quantify levels of RNA species in differing cell types, or to test the functional importance of a sequence or secondary structure in an RNA molecule of interest. Thus, experimental tools that accomplish these goals have widespread usefulness in diverse areas of molecular biology, biochemistry, and cell biology.

The purpose of this chapter is to present a concise series of experimental procedures that can be used for the rapid and decisive characterization of RNA. Due to recent advances in studies of nuclear pre-mRNA splicing, including remarkable growth in the identification of alternatively spliced genes, methods will be emphasized that are particularly useful in these areas. In addition, this chapter will address problems encountered in the detection of small sequence changes in otherwise long RNA chains, and the measurement of low abundance RNA species.

2. Isolation of RNA

RNA can be isolated by any of the procedures described elsewhere in this book (see Chapter 1). Successful application of the methods described in this chapter requires that the RNA starting material is of high quality. Therefore, the following precautionary steps are recommended in order to avoid degradation of RNA:

(a) Take steps to inactivate RNases in both the biological starting material and the solutions and apparatus used (see Chapter 1, Section 1.1).

(b) Store RNA samples on ice whenever possible during analysis; otherwise store RNA at $-70\,°C$.

(c) Use water that has been distilled and deionized and then sterilized to prepare reaction mixtures.

(d) Dismantle and decontaminate automatic pipetting devices frequently according to the manufacturer's instructions.

(e) Treat Plexiglass electrophoresis tanks, gel combs, and spacers with Chlorox bleach to remove any contaminating nucleases prior to use in gel electrophoresis of RNA samples.

3. Procedures for establishing the identity of spliced RNAs

3.1 Introduction

Identification of RNAs that are alternatively spliced depends upon distinguishing RNA species that, by definition, share common sequences and are frequently similar in length. These RNAs can be identified by nuclease S1 analysis using a cDNA probe. Results obtained with one nuclease S1 probe should be confirmed with a second probe, or by another method such as Northern blot analysis with exon- and intron-specific oligonucleotide probes. Obtaining confirmatory evidence can be cumbersome if suitable cDNA probes are not available for each of the predicted spliced RNAs, or if oligonucleotide probes give a low efficiency of hybridization in the Northern blot analysis. A two-step method which combines the activities of RNase H and nuclease S1 is a rapid, alternative approach. In addition, if the RNA to be identified is present in low amounts, polymerase chain reaction (PCR) technology will be required in conjunction with cDNA synthesis by reverse transcriptase (RT–PCR, or cDNA–PCR) to amplify the RNA sequence sufficiently for detection.

3.2 Basic nuclease S1 analysis

3.2.1 Strategy

Nuclease S1 analysis is a highly sensitive, quantitative tool used to

- identify spliced RNAs based on sequence colinearity with a complementary DNA probe
- map the positions of splice junctions, polyadenylation sites, and 5′ ends of RNA molecules

The identification of spliced RNA is accomplished by hybridization of a cDNA probe, labelled at its 5′ or 3′ end, to the RNA of interest followed by treatment with nuclease S1, which specifically degrades single-stranded regions (1,2). The correctly spliced RNA will protect a specific sequence of the DNA fragment, whose length can be measured by electrophoresis in a denaturing polyacrylamide gel with appropriate molecular weight markers. Consistent results should be obtained with both 5′ and 3′ end-labelled probes.

Mapping of the splice sites can be accomplished by using end-labelled DNA probes containing sequences complementary to both the exon and adjacent intron regions that span the splice site. In this case, the spliced RNA will protect only the exon-containing sequences of the probe up to the position of the splice site. To map the 5' splice site junction, the radioactive label must be at the 3' end of the DNA probe. To map the 3' splice site junction, use a 5' end-labelled probe. A similar strategy can be used to map the 5' end and the polyadenylation site of a primary transcript. A typical nuclease S1 analysis of spliced RNA, using a 3' end-labelled probe, is illustrated in *Figure 1*.

3.2.2 Preparation of 5' and 3' end-labelled probes

DNA probes suitable for nuclease S1 analysis are normally prepared by digestion of the appropriate cDNA plasmids with restriction enzymes, or by enzymatic extension of synthetic oligonucleotides (3).

i. 3' end-labelling

The preparation of recombinant plasmids is described in Chapter 1, Section 2.1. Labelling the DNA probe at the 3' end is carried out by linearizing the plasmid with a restriction enzyme that generates a 3' recessed end at the site to be labelled. Labelling of the 3' ends of the DNA fragment is performed in the restriction enzyme reaction by addition of DNA polymerase I (Klenow fragment) and the appropriately labelled [α-^{32}P]deoxynucleoside triphosphate. A second restriction enzyme is then used to remove the label from the end of the DNA fragment that is distal to the probe sequence. This procedure is described in *Protocol 1* where the two enzymes *Eco*RI and *Hin*dIII are used for illustrative purposes. When using other restriction enzymes, different labelled and unlabelled deoxyribonucleoside triphosphates may have to be used as required for the particular 3' recessed end involved.

Protocol 1. Preparation of 3' end-labelled probes: Klenow labelling of 3' recessed ends

Reagents and equipment

- TE buffer (10 mM Tns–HCl, pH 8.0, 1 mM EDTA)
- DNA plasmid which contains sequence to be labelled (0.25 µg in TE buffer).
- Restriction enzymes, *Eco*R1 and *Hin*dIII (New England Biolabs, ~20 units/µl)[a]
- 10 × restriction buffers for *Eco*R1 and *Hin*dIII (compositions as recommended by the enzyme supplier)
- 10 × Klenow buffer (0.1 M Tris–HCl, pH 7.5, 50 µM MgCl$_2$, 75 mM DTT)
- Klenow enzyme (Klenow fragment of DNA polymerase I New England Biolabs, 5000 units/ml)
- 1% agarose gel and electrophoresis equipment (see Chapter 1, *Protocol 2*)
- 5 mM dTTP[b]
- [α-^{32}P]dATP (Amersham; ~3000 Ci/mmol, 10 µCi/ml)[b]
- 3 M sodium acetate, pH 4.0
- 0.5 M EDTA
- G50-150 Sephadex spin column (3 ml size)

Continued

2: Characterization of RNA

Protocol 1. Continued

Method

1. Combine 0.25 µg of the DNA plasmid, 2.5 µl of 10× restriction buffer, 1 µl of restriction enzyme in a microcentrifuge tube with H_2O to a final volume of 25 µl. Choose an appropriate restriction enzyme to linearize the DNA at a site located at the 5' end of the insert to be labelled, for example *Eco*R1. The 5'-end of the insert refers to the 5'-end of the sense strand of the DNA. The restriction enzyme must produce 3'-recessed ends.
2. Incubate the restriction digestion reaction at 37 °C for $\geqslant 1$ h to digest the DNA sample completely. After 1 h, remove a small aliquot of the restriction digestion reaction and analyse it by electrophoresis on a 1% agarose gel to check for complete digestion (see Chapter 1, *Protocol 6*).
3. Add to the remainder of the restriction digestion reaction in step 2, 2 µl of 5 mM dTTP, 4 µl of 10× Klenow buffer, 5 µl of $[\alpha\text{-}^{32}P]$dATP, 1 µl of Klenow ezyme, and H_2O to a final volume of 40 µl.
4. Incubate the mixture at 30 °C for 15 min.
5. Stop the reaction by the addition of 0.5 M EDTA to a final concentration of 10 mM.
6. Load the entire sample on to a G50-150 Sephadex spin column, which has been pre-equilibrated in TE buffer, and centrifuge for 4 min at 2000 r.p.m. Collect the eluate containing the DNA (excluded volume).
7. Concentrate the labelled DNA fragment by ethanol precipitation as described in *Protocol 3*, step 1.
8. Resuspend the DNA in 10 µl of TE buffer and digest it as in step 2 with a second restriction enzyme, for example *Hind*III, at a site just upstream of the *Eco*R1 site. This digestion will remove the label from the *Eco*R1 end that is distal to the insert and preserve the label located at the end closer to the insert. The label will be located on the DNA strand that is complementary to the RNA species to be analysed.
9. Concentrate the DNA sample, if required, by ethanol precipitation as in step 7.

[a] Other enzymes should be chosen to suit other plasmids.
[b] These dNTPs are appropriate for *Eco*R1. Substitute others as required for the site involved.

ii. 5' end-labelling

Labelling the cDNA strand at the 5'-end can be achieved by digesting the plasmid with an appropriate restriction enzyme, followed by phosphatase treatment, then phosphorylation using T4 polynucleotide kinase (PNK) and $[\gamma\text{-}^{32}P]$ATP. A second restriction enzyme is then used to remove the label from the end of the DNA fragment that is distal to the probe sequence. Alternatively, synthetic oligonucleotides can be used as probes. In this instance the procedure is to phosphorylate the 5'-ends in one step with polynucleotide kinase as described in *Protocol 2*.

Protocol 2. Preparation of 5′ end-labelled oligonucleotide probes

Reagents

- Oligonucleotide stock solution (10 ng/µl)
- 5× kinase buffer (0.25 M Tris–HCl, pH 9.0, 50 mM MgCl$_2$, 25% glycerol)
- 0.1 M DTT
- [γ-^{32}P]ATP (Amersham; ~3000 Ci/mmol, 10 mCi/ml)
- T4 polynucleotide kinase (New England Biolabs; 10 units/µl)
- 0.5 M EDTA
- Centrifuge tubes (Falcon 2063)
- TE buffer (10 mM Tris–HCl, pH 8.0, 1 mM EDTA)
- Bio-Gel P6 beads (BioRad), hydrate beads overnight in sterile H$_2$O, rinse them twice in TE buffer and store them at 4 °C, Bio-Gel P6 has an exclusion limit of 6000 Da, this gel-filtration reagent can be used for oligonucleotides as well as DNA fragments in the procedure described below.

Method

1. Combine 5 µl of oligonucleotide stock solution, 5 µl of 5× kinase buffer, 1.25 µl of 0.1 M DTT, 5 µl of [γ-^{32}P]ATP, and 1 µl of PNK in a microcentrifuge tube with sterile H$_2$O to a final volume of 25 µl. For each pmol of oligonucleotide present in the reaction add one pmol of [γ-^{32}P]ATP. If a double-stranded DNA fragment is used as the substrate, use 2 pmol of [γ-^{32}P]ATP for each pmol of DNA.
2. Incubate the reaction at 37 °C for 30 min.
3. Add 2 µl of 0.5 M EDTA to terminate the reaction.
4. Set up the Biol-Gel P6 column. Poke a hole in the bottom of a 0.5 ml microcentrifuge tube using a 25 gauge needle, then place this tube inside a 1.5 ml microcentrifuge tube. Layer 1–2 mm of glass beads (rinsed in TE buffer) on the bottom of the inner tube, then add sufficient Bio-Gel P6 beads to fill the tube to approximately 80% of capacity. Place the microcentrifuge tubes inside a Falcon tube as a holder and centrifuge the tubes at 1000 g for 1 min to pack the column. Transfer the tube containing the packed P6 column to a fresh 1.5 ml microcentrifuge tube.
5. Apply the radioactive sample from step 3 and centrifuge at 1500 g for 5 min. Retrieve the labelled oligonucleotide from the solution at the bottom of the 1.5 ml microcentrifuge tube (column flow-through) by ethanol precipitation (see *Protocol 3*, step 1). Dispose of the P6 column, which contains unreacted [^{32}P]ATP, using appropriate precautions.

3.2.3 Nuclease S1 procedure

Protocol 3 describes the basic technique of nuclease S1 analysis.

2: Characterization of RNA

Protocol 3. Nuclease S1 analysis: the basic technique

Reagents

- 8× hybridization buffer (0.32 M Pipes–KOH, pH 6.4, 8 mM EDTA, 3.2 M NaCl)
- Deionized formamide[a]
- DNA probe, ^{32}P-end-labelled (see *Protocols 1* and *2*, 1.4 ng/µl H$_2$O)
- Nuclease S1 (Sigma; specific activity, 200 000–600 000 units/mg)
- Nuclease S1 buffer (0.28 M NaCl, 50 mM sodium acetate, pH 4.6, 4.5 mM ZnSO$_4$, 20 µg/ml salmon sperm DNA). Denature the salmon sperm DNA just before use by boiling it for 10 min in a microcentrifuge tube with a small hole punctured in the lid. Then dilute this immediately into the nuclease S1 buffer[b]
- Stop buffer (4.0 M NH$_4$OH, 0.1 M EDTA)
- Glycogen (molecular biology grade; Calbiochem) or tRNA (*E. coli*; Sigma) carrier
- HHS buffer (20 mM Hepes–KOH, pH 7.5, 0.25 M sodium acetate, 0.1 mM EDTA)
- Phenol saturated with HHS buffer
- TBE buffer (50 mM Tris base, 50 mM boric acid, 50 µM EDTA)
- Formamide–dye mixture: make this by mixing 95% deionized formamide with 5% dye mix (20% sucrose, 0.13 g bromophenol blue, 0.13% xylene cyanol, 0.1× TBE, 0.1% SDS)
- 8% polyacrylamide/7 M urea gel (see Chapter 1, *Protocol 8*) and appropriate DNA molecular weight markers
- 3 M sodium acetate, pH 4.0

Method

1. For each reaction to be performed, ethanol precipitate 20 µg of the RNA sample plus 21 ng of DNA probe in a sterile microcentrifuge tube by adding 0.1 vol. of 3 M sodium acetate, pH 4.0, and 2.5 vol. of absolute ethanol. Mix by vortexing, incubate at −70 °C for 1 h and then collect the nucleic acid precipitate by centrifugation for 15 min in a microcentrifuge. Remove the supernatant and vacuum dry the RNA pellet.

2. Resuspend the dried pellet (RNA plus DNA probe) by adding the following reagents then leave for 10 min on ice:

   - 8× hybridization buffer   1.9 µl
   - deionized formamide   12.0 µl
   - H$_2$O   1.1 µl

3. Denature the solution of RNA plus probe (mixture from step 2) at 80 °C for 15 min.

4. Hybridize for 3 h at the appropriate temperature[c].

5. Prepare the nuclease S1 mixture by adding 800 units of nuclease S1 to 150 µl of ice-cold nuclease S1 buffer for each reaction to be performed[d].

6. Add 150 µl of the ice-cold nuclease S1 mixture to each tube from step 4 and incubate at 37 °C for 30 min.

7. Stop each reaction by adding 50 µl of ice-cold stop buffer.

Continued

Protocol 3 *Continued*

8. Adjust the volume to 0.4 ml with HHS buffer and extract by vortexing with an equal vol. of HHS-saturated phenol. Separate the phases by centrifugation in a microcentrifuge tube. Carefully remove the upper (aqueous) phase.
9. Extract the aqueous phase with an equal volume of chloroform.
10. Add 10 µg/ml of glycogen or tRNA carrier to the aqueous phase and ethanol precipitate the nucleic acids as in step 1.
11. Following the nuclease S1 procedure, resuspend the dried nucleic acid pellet in 10 µl of formamide–dye mixture, and denature the hybrid by boiling for 5 min just prior to electrophoresis. Resolve DNA fragments by electrophoresis in a denaturing polyacrylamide gel as described in Chapter 1, *Protocol 8*. An 8% polyacrylamide/7 M urea gel is suitable for resolving fragments ranging in length from approximately 100–900 nt. Co-electrophorese DNA molecular weight markers or a dideoxynucleotide sequencing ladder alongside to enable the protected fragments to be sized.

[a] Deionize the formamide by stirring it with a mixed-bed ion-exchange resin (e.g. Amberlite MB-3, Sigma). Store it in aliquots at $-20\ °C$.
[b] A sample of salmon sperm DNA that has not been fully denatured immediately before use may result in insufficient single-stranded DNA carrier in the reaction.
[c] Hybridization is routinely performed at 15–25 °C below the T_m of the hybrid. The T_m can be calculated from the following equation (4) for probes of length 100 nt or greater:

$$T_m\ °C = 81.5 - 16.6(\log[Na^+]) + 0.41(\%G+C) - 0.63(\%\ formamide) - (600/l)$$

where l is the length of the expected hybrid in base pairs and (G+C) is the proportion of guanosine plus cytidine residues in the probe.
[d] It is essential that the correct ratio of probe and nuclease S1 is used. Carry out a pilot experiment as described in Section 3.2.4.

3.2.4 Control of nuclease S1 activity

The success of the nucleases S1 assay is critically dependent upon an appropriate ratio of nuclease S1 to probe. Each batch of nuclease S1 should be titrated with a constant amount of cDNA probe (21 ng) and carrier DNA (20 mg/ml) under the conditions of the assay (see *Protocol 3*). This titration will allow a precise determination of the amount of enzyme needed to digest completely the unprotected (single-stranded) cDNA probe. Too little enzyme will obviously result in incomplete digestion of the probe, which will minimize the signal from the protected fragments. In some cases, protected fragments may even be overshadowed by the undigested probe. On the other hand, too much enzyme will result in overdigestion of the protected fragments leading to inaccurate and misleading estimates of the protected sequence and/or 'nicking' of the probe in AT- or AU-rich regions of the hybrids. In the most severe cases, no protected fragments at all will be recovered.

3.2.5 Potential problems and suggested remedies

i. No protected fragments are recovered
A variety of factors can contribute to the lack of recovery of protected DNA fragments. As a first step, the integrity of the input RNA sample should be evaluated by electrophoresis of a small aliquot (0.5–1.0 µg RNA) on a 1% agarose gel in standard buffer (see Chapter 1, *Protocol 6*). This check is especially recommended for RNA samples derived from cells or tissues. Denature the RNA sample briefly (5 min at 65 °C) in formamide–dye mixture (see *Protocol 3*) just prior to electrophoresis. Because of their abundance, the ribosomal RNAs in the sample serve as diagnostic indicators of the quality of the RNA sample as a whole. If the ribosomal RNAs appear degraded, the preparation of a new RNA sample is warranted.

ii. Protected fragments are recovered with only one of the end-labelled probes
Some DNA fragments chosen for use as probes in nuclease S1 analysis may not be compatible with the assay due to loss of the label during the nuclease S1 digestion. This may be due to RNA:DNA duplex formation that is insufficiently stable, or to an interfering secondary structure in the vicinity of the label. Selection of a longer or shorter version of the same cDNA probe is a possible remedy.

iii. Inconsistent recovery of protected fragments
This may be due to contamination of the electrophoresis apparatus and/or pipetting devices with nuclease S1. These can be decontaminated as suggested in Section 2.

3.3 Combined RNase H and nuclease S1 analysis

This method is particularly useful in distinguishing spliced RNAs that differ by one or more short, internal sequences.

(a) In the first step, the RNA sample is hybridized to an oligodeoxyribonucleotide complementary to a sequence within an individual exon and then treated with RNase H. The procedure is described in *Protocol 4*. RNase H will cleave only the RNA strand in the specific RNA:DNA hybrid, leaving the remaining RNA segments intact (5).

(b) In the second step, the resulting RNA is subjected to nuclease S1 analysis with a single cDNA probe, as described in *Protocol 3*.

Protocol 4. RNase H treatment prior to nuclease S1 analysis

Reagents

- 10 × RNase H buffer (1.0 M KCl, 0.1 M MgCl$_2$, 1 mM EDTA, 0.2 M Hepes–KOH, pH 7.4)
- RNase H (Promega; 800 units/ml)

Continued

Protocol 4 *Continued*

- Oligonucleotide stock solutions (0.25 µg/µl) prepared in sterile water), oligonucleotides used here are typically 18 nt in length
- RNA sample, HHS buffer, stop buffer, glycogen (or tRNA) as in *Protocol 3*
- Phenol saturated with HHS buffer
- Chloroform

Method

1. Combine the reaction components in a sterile microcentrifuge tube, at 4 °C, as follows:
   - RNA sample, 10 µg/µl   2 µl
   - 10 × RNase H buffer   1 µl
   - oligonucleotide (0.25 µg/µl) 2 µl
   - H_2O 4 µl
2. Denature the nucleic acids by incubating the sample mixture at 80 °C for 3 min.
3. Hybridize for 30 min at an appropriate temperature as described in *Protocol 3*, footnote c.
4. Cool the hybridization mixture to room temperature and add 1 µl of RNase H.
5. Incubate at 30 °C for 1 h.
6. Extract successively with saturated phenol and then with chloroform. Ethanol precipitate the nucleic acids as described in *Protocol 3*, step 1.
7. Carry out the nuclease S1 analysis as in *Protocol 3*.

The autoradiogram of *Figure 1* shows an example of nuclease S1 analysis which includes the RNase H treatment (*Protocol 4*). In this experiment, two alternatively spliced RNAs are present in a sample of RNA isolated from HeLa cells transfected with a rat preprotachykinin gene construction. Where nuclease S1 analysis was preceded by treatment with RNase H in the presence of a non-specific oligonucleotide (lane B), there are two protected fragments, 222 nt and 153 nt in length. (Note: this result is equivalent to the standard nuclease S1 treatment described in *Protocol 3*). When the RNA was first hybridized to an exon 3-specific oligonucleotide, treated with RNase H and then nuclease S1 (lane C), both fragments were decreased in size to 97 nt. In contrast, when an exon 4-specific oligonucleotide was used (lane D), only the 222 nt band is decreased to 173 nt in size (see *Figure 1* schematic diagram). Thus, the RNase H treatment verifies that only the 222 nt protected fragment arises from an RNA transcript containing exon 4 sequences. Alternatively, to confirm the results obtained using one DNA probe labelled at the 3' end, use a second probe, which is labelled at the 5' end. The advantage of the combined RNase H and nuclease S1 procedure is that it provides a rapid method for the analysis of spliced RNA products using a single nuclease S1 probe.

2: Characterization of RNA

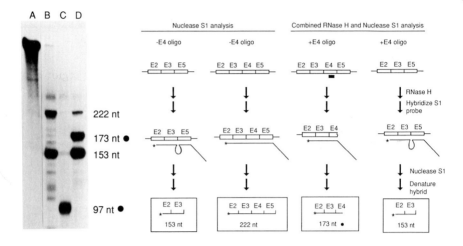

Figure 1. Nuclease S1 analysis of alternatively spliced RNA products after RNase H treatment. Total RNA was isolated from HeLa cells after transfection with a segment of the rat preprotachykinin gene extending from exon 2 to exon 5 (ref. 6). A schematic diagram of the pre-mRNA is shown (top). The left-hand side of the figure shows the autoradiogram of the results; the right-hand side illustrates schematically the reactions involved and the protected products. **Autoradiogram** (left): Lane A, intact, 3' end-labelled probe corresponding to an 884 bp BanI–HindIII segment of the cDNA, containing exons 2, 3, 4, 5, 6 and 7. Step 1: each RNA sample (20 µg) was treated with RNase H and a non-specific oligonucleotide (lane B), an exon 3-specific oligonucleotide (lane C), or an exon 4-specific oligonucleotide (lane D). The exon 3-specific oligonucleotide is complementary to nucleotides 42–61 of exon 3, whereas, the exon 4-specific oligonucleotide is complementary to nucleotides 21–39 of exon 4. Step 2: RNA samples were hybridized to the S1 probe. Step 3: RNA samples were treated with nuclease S1. Step 4: RNA:DNA hybrids were heat denatured. Filled circles mark protected fragments whose mobilities have been altered by specific RNase cleavage. **Schematic diagram** (right): Illustration of the reactions of lanes B and D. RNA spliced products are indicated by the fused boxes (exons). The protected S1 probes, 3' end-labelled (asterisk), are boxed (bottom). Note: the 173 nt product shown in lane D contains the following sequences: exon 2 (56 nt), exon 3 (97 nt) and the 5' portion of exon 4 (20 nt). The 153 nt product, also shown in lane D, lacks exon 4 sequences, but is otherwise identical to the 173 nt product

3.4 RT–PCR analysis

3.4.1 Introduction

Characterization of spliced RNAs by a combination of cDNA synthesis with reverse transcriptase followed by PCR amplification (7) is particularly useful for RNA products that are present in low abundance or that contain stable secondary structure(s). Because of the PCR amplification, this can be an extremely sensitive method for detecting spliced RNA molecules.

3.4.2 cDNA synthesis from an RNA template

The cDNA strand is synthesized from the RNA template using reverse transcriptase. AMV reverse transcriptase (BioRad), or the manganese-dependent reverse transcriptase activity contained in the enzyme rTth DNA polymerase (Perkin Elmer Cetus; refs 8, 9) are recommended. The rTth DNA polymerase is thermostable so that cDNA synthesis can be carried out at an elevated temperature, which destabilizes RNA secondary structures. Both enzymes require a primer which should be an oligonucleotide complementary to a region of the RNA template 3' to the portion of the RNA under investigation. Label this downstream primer at its 5' end with ^{32}P as described in *Protocol 2*. The synthesis of the cDNA is described in *Protocol 5*.

Protocol 5. Synthesis of cDNA from an RNA template

Reagents

- RNA template (<125 ng/µl) prepared by one of the methods described in Chapter 1, Section 3
- 10×RT buffer (0.9 M KCl, 0.1 M Tris–HCl, pH 8.3)
- 10 mM MnCl$_2$
- dNTP stock solutions (separate 10 mM stock solutions of dATP, dCTP, dGTP, dTTP)
- rTth DNA polymerase (Perkin Elmer Cetus)
- Downstream primer (15 µM, equivalent to 100 ng/µl for a 20 nt oligonucleotide)

Method

1. Combine the following components for each reaction in a microcentrifuge tube:
   - RNA template   2 µl
   - 10×RT buffer   2 µl
   - downstream primer   1 µl
   - 10 mM MnCl$_2$ 2 µl
   - dNTP stock solutions   0.4 µl each
   - rTth DNA polymerase   5 units
   - sterile H$_2$O   to 20 µl
2. Overlay the contents of each tube with 100 µl of mineral oil.
3. Incubate for 5–15 min at 70 °C.
4. Stop the reactions by removing the tubes to an ice bucket.
5. Amplify the cDNA by PCR as described in *Protocol 6* (Section 3.4.3).

3.4.3 PCR amplification of cDNA

Amplification of the cDNA first strand uses a standard PCR procedure described in *Protocol 6*. A second or upstream primer is required. This should be an oligonucleotide complementary to the cDNA strand and located upstream of the region to be amplified. The upstream primer is routinely 5' end-labelled (see *Protocol 2*).

2: Characterization of RNA

Protocol 6. PCR amplification of cDNA

Equipment and reagents

- 10 × PCR buffer (0.1 M Tris–HCl, pH 8.3, 1 mM KCl, 7.5 mM EGTA, 0.5% Tween 20, 50% glycerol)
- 25 mM $MgCl_2$
- ^{32}P-labelled upstream primer (1 µl of 15 µM, equivalent to 100 ng/µl of a 20 nt oligonucleotide; 25% of the molecules are ^{32}P-labelled, see *Protocol 2*)
- cDNA synthesized as in *Protocol 5*
- PCR thermocycler apparatus
- Materials for polyacrylamide/urea gel electrophoresis as described in Chapter 1, *Protocol 8*

Method

1. For each PCR reaction, mix the following reagents in a microcentrifuge tube:
   - 10 × PCR buffer 8 µl
   - 25 mM $MgCl_2$ 6 µl
   - ^{32}P-labelled upstream primer 1 µl
   - sterile H_2O to 80 µl
2. Add the mixture to the cDNA synthesis reaction from *Protocol 5*, step 4.
3. Mix by gentle vortexing. Centrifuge briefly in a microcentrifuge. This removes air bubbles, returns the contents to the bottom of the tube, and restores the mineral oil overlay.
4. Place the tubes in a PCR thermocycler and preincubate at 95 °C for 2 min.
5. Amplify the cDNA through 15–25 cycles consisting of 95 °C, 1 min; 65 °C, 1.5 min; 72 °C, 2.5 min.
6. Incubate at 72 °C for 10 min for the final extension reaction.
7. Transfer to 4 °C and remove the mineral oil by extracting with an equal volume of chloroform.
8. Reduce the volume to approximately 2 µl by lyophilization in a Speedvac (approx. 1 h).
9. Separate the PCR products in a denaturing polyacrylamide/urea gel as described in Chapter 1, *Protocol 8*.

3.4.4 Anticipated results and possible pitfalls

The experiment shown in *Figure 2*, is an example of RT–PCR and suggests how pitfalls in the technique can be prevented (described below). This experiment used total RNA from HeLa cells that had been transfected with a rat preprotachykinin gene construction. The gene construction contains exons 2, 3, 4 and 5, with associated introns. Expression of this gene construction results in two alternatively spliced RNAs, one containing exons 2 + 3 + 4 + 5, and one containing exons 2 + 3 + 5.

Besides illustrating results typical of an RT–PCR analysis, three features of the experiment in *Figure 2* indicate how pitfalls in the technique can be prevented.

(a) In order to measure the ratio of two spliced RNAs accurately it is important to establish that DNA amplification was performed in the exponential phase of the PCR. *Figure 2* shows three identical RT–PCR reactions, which were terminated after 20, 23, and 26 cycles of amplification, respectively (lanes A, B, and C). Reactions shown in lanes A and B were terminated during the exponential phase of the PCR (exponential accumulation of products), whereas the reaction shown in lane C was not. To determine conditions for exponential amplification, it is necessary to test different amounts of input RNA over a range of PCR cycles.

(b) It is important to verify that sample cross-contamination does not contribute to the PCR products. A mock amplification reaction is a sensitive measure of cross-contamination. As expected for a reaction containing all components except input RNA, no PCR products were observed in the reaction shown in lane D of *Figure 2*. Since pipetting devices can be an important source of this type of contamination, aerosol-resistant pipette tips should be used for all steps of the experiment until the amplified products are obtained.

(c) Because of the enormous sensitivity of PCR amplification it is essential to ensure that the RNA sample was not contaminated by genomic DNA isolated along with the RNA from the original biological source. RNase sensitivity of the input RNA sample is a critical test to discount this type of contamination. The reaction shown in lane E of *Figure 2* demonstrates that the PCR amplification was RNase-sensitive as judged by the loss of both PCR products. RNA samples for analysis by RT–PCR may be purified free of contaminating DNA by extensive incubation with RNase-free DNase (Promega).

3.5 Primer extension analysis of RNA

3.5.1 Introduction

The strategy of the primer extension procedure is to hybridize a ^{32}P-labelled oligonucleotide (primer) to the RNA transcript followed by the addition of reverse transcriptase and deoxynucleoside triphosphates. Reverse transcriptase will then synthesize a DNA strand that is complementary to the RNA template, beginning at the primer. The primer extension will stop at the 5' end of the RNA strand, or alternatively at structural impediments such as secondary structure or branchpoints (see below). Several examples of primer extension are illustrated schematically in *Figure 3*.

The primer extension mapping procedure is versatile for several reasons:

(a) Intron- or exon-specific primers can be used to discriminate spliced products from intron-containing pre-mRNAs or intermediates.

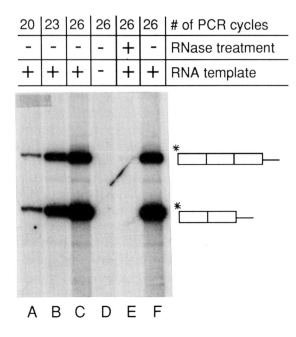

Figure 2. RT-PCR analysis of alternatively-spliced RNA. HeLa cells were transfected with a rat preprotachykinin gene construction consisting of exons 2-5. Total RNA was isolated and reverse transcribed using an unlabelled downstream primer. The downstream primer is complementary to IVS5 sequences (located downstream of exon 5). The cDNA first strand was then amplified by addition of an upstream primer, labelled at its 5' end, through 20, 23, or 26 cycles of PCR (lanes A, B, C, F). The upstream primer is the same sequence as the first 19 nt of exon 3 (RNA). The PCR products were separated on a 10% polyacrylamide gel and autoradiographed. The PCR products derived from alternatively spliced RNAs containing exons 3 + 5 and 3 + 4 + 5 are indicated to the right of the autoradiogram (bottom and top, respectively). Controls were included where the original RNA template was omitted (lane D) or treated with RNase A (lane E). Boxes, exon sequences; lines, intron sequences

(b) The sequence to which the primer is hybridized can be varied to obtain short or long primer extension products.

(c) To discriminate internal sequence differences in closely-related spliced RNAs, primer extension analysis can be combined with RNase H treatment.

3.5.2 Basic procedure

Protocol 7 describes the basic method for primer extension.

Protocol 7. Primer extension analysis

Reagents

- 10 × hybridization buffer (3.0 M NaCl, 0.1 M Tris–HCl, pH 7.5, 10 mM EDTA)
- 10 × extension buffer (0.6 M NaCl, 0.1 M Tris–HCl, pH 8.1, 500 µg/ml actinomycin D, 80 mM $MgCl_2$; store this solution in aliquots at −20 °C)
- Oligonucleotide primer, ^{32}P-labelled at its 5′ end as in *Protocol 1* (0.5 ng/µl H_2O)
- 1.0 M DTT
- Separate 0.1 M stock solutions of dATP, dCTP, dGTP, dTTP
- AMV reverse transcriptase (BioRad; 20 000 units/ml)
- RNA sample, HHS buffer, and reagents for phenol extraction and ethanol precipitation of primer extension products as in *Protocol 3*
- 10% polyacrylamide/7.0 M urea sequencing gel, formamide–dye mixture, and size markers (see Chapter 1, *Protocol 8*)

Method

1. Prepare the extension reaction mixture by combining the following reagents (sufficient for 20 primer extension reactions):

10 × extension buffer	66 µl
1.0 M DTT	6.6 µl
0.1 M dATP	6.6 µl
0.1 M dCTP	6.6 µl
0.1 M dGTP	6.6 µl
0.1 M dTTP	6.6 µl
reverse transcriptase	5 units/µl final concentration
sterile H_2O	to 466 µl final volume

2. For each extension reaction combine the following reagents in a microcentrifuge tube:

● RNA (1–10 µg)	8 µl
● 10 × hybridization buffer	1 µl
● ^{32}P-labelled primer (0.5 ng)	1 µl

3. Mix by vortexing gently. Hybridize for 10 min at an appropriate temperature. The hybridization temperature is normally in the range of 50–70 °C, but it is best to determine the optimum temperature, empirically, for each primer. The T_m of the hybrid can be estimated by the equation given in *Protocol 3* (footnote c) for primers in the range of 14–70 nt. Hybridization should be performed at a temperature that is 5–10 °C below the T_m of the hybrid.

4. Transfer the tubes *directly* to 42 °C and add 23.3 µl of pre-warmed (42 °C) extension reaction mixture from step 1[a]. Pipette up and down to mix. Allow the extension reaction to proceed for 20 min at 42 °C.

Continued

2: Characterization of RNA

Protocol 7. Continued

5. Transfer the tubes to an ice bucket and immediately adjust the volume of each reaction to 400 µl with HHS buffer.
6. Extract each reaction successively with phenol and chloroform and then ethanol precipitate the extension products as in *Protocol 3*, step 1.
7. As with the nuclease S1 analysis, primer extension products are separated by denaturing polyacrylamide gel electrophoresis and then visualized by autoradiography. Resolve primer extension products next to DNA sequencing ladders on a 10% polyacrylamide sequencing gel after denaturing samples in 95% formamide–dye mixture for 5 min at 95 °C as described in Chapter 1, *Protocol 8*.

[a] High background levels may be observed if samples are cooled even briefly between the hybridization and extension reaction steps.

3.5.3 Limits of detection

The sensitivity of the primer extension technique can be quantified by testing a dilution series of an RNA molecule of known molecular weight and starting concentration. A test RNA of 1000 nt at 0.1–1.0 ng should give a strong autoradiographic signal after a 2 h exposure. Even less RNA (25 pg) can easily be detected after an overnight exposure.

3.6 Northern blot analysis with exon and intron probes

3.6.1 Introduction

Northern blot analysis is a highly sensitive method for the detection of full-length RNA immobilized on a nylon membrane. Because it is direct, this method is useful for the detection of spliced RNA derived from cultured cells, tissues, or *in vitro* splicing reactions. The limit of detection of an RNA molecule 1000 nt in length is approximately 10 pg. However, this sensitivity depends strongly upon the efficiency of RNA transfer, the specific activity of the probe, and the efficiency of hybridization. This method is frequently used in combination with a high resolution mapping technique such as nuclease S1 or primer extension analysis (see Sections 3.2 and 3.5).

Northern blot analysis is useful for sorting out complex splicing patterns since:

(a) The full size range of RNA intermediates and products can normally be resolved on one or two denaturing polyacrylamide gels of different concentrations (10).
(b) Probes complementary to individual exons and introns can be used to establish the identity of RNA species by size and sequence.
(c) The existence of branchpoint-containing intermediates (lariats) can be detected and later confirmed by subsequent branchpoint analysis (see Section 4).

3.6.2 Basic procedure

Northern blotting consists of three principal steps:

- resolution of the RNA sample by denaturing gel electrophoresis
- transfer of the resolved RNA species to a nylon membrane
- hybridization of the transferred RNA to radiolabelled probe(s)

The choice of gel used to perform the initial separation depends primarily on the range of RNA species to be separated; 4–10% polyacrylamide/7.0 M urea gels normally suffice for the separation of RNAs up to approximately 1400 nt. Agarose gels are recommended for much larger RNAs. Electrophoretic separation of RNA samples prior to Northern blot analysis is described in Chapter 1 (Section 2.3). *Protocol 8* gives a procedure for the transfer of the RNA after electrophoresis and *Protocol 9* describes the hybridization analysis.

Protocol 8. Transfer of RNA to nylon membranes

Equipment and reagents

- UV transilluminator (Stratalinker; Stratagene)
- Transfer buffer (0.1 M Tris–HCl, pH 7.8, 50 mM sodium acetate, 5 mM EDTA)
- 6 × SSC buffer (0.9 M NaCl, 90 mM trisodium citrate)
- Nylon transfer membrane, GeneScreen membrane (DuPont NEN) is particularly suitable.
- Electroblotting apparatus (e.g. Transblot Cell, BioRad)
- Gel containing electrophoresed RNA sample and radioactive molecular weight markers (e.g. pBR322 *Msp*1 digest)

Method

NB Wear plastic or latex disposable gloves when handling the nylon membrane and gel to prevent transfer of nucleic acids from the hands.

1. Soak the gel for 30 min in the buffer used for electrophoresis. This step removes the urea from the gel.
2. Equilibrate the gel for 10 min in transfer buffer.
3. Cut the nylon membrane to fit the gel. Soak it in transfer buffer.
4. Place the nylon membrane in contact with the equilibrated gel and insert both into the electroblotting apparatus, following the manufacturer's instructions.
5. Electroblot for >4 h at 200 mA. Cool the apparatus to 4 °C while the transfer is in progress[a].
6. Open the apparatus and remove the nylon membrane.
7. Rinse the blot in 6 × SSC buffer.

Continued

2: Characterization of RNA

Protocol 8. *Continued*

8. Immobilize the RNA on to the membrane using UV light. Keep the blot damp, but not excessively wet, during the procedure.
9. Either hybridize the blot immediately with the radioactive probe (see *Protocol 9*) or store it (at 4 °C protected from light) for hybridization later.

[a] The efficiency of transfer of the RNA from the gel to the nylon membrane can be monitored conveniently by using the radioactive size markers co-electrophoresed with the RNA samples.

Protocol 9. Hybridization of Northern blots

Reagents

- Prewash buffer (50 mM Tris–HCl, pH 8.0, 1.0 M NaCl, 1 mM EDTA, 0.1% SDS)
- 50 × Denhardt's solution: dissolve 1 g Ficoll, 1 g polyvinylpyrrolidone, 1 g BSA (molecular biology grade) in sterile H_2O to 100 ml final volume; filter sterilize using 0.45 micron filters and store in aliquots at −20 °C
- 20 × SSPE buffer: add 174 g NaCl, 27.6 g $NaH_2PO_4 \cdot H_2O$ and 7.4 g disodium EDTA to just under 1 litre of H_2O; stir overnight to dissolve; add H_2O to 1 litre final volume
- 10% SDS
- 20 × SSC buffer (3.0 M NaCl, 0.3 M trisodium citrate)
- Salmon sperm DNA (10 mg/ml) dissolved in H_2O
- ^{32}P-labelled DNA probe (2–5 × 10^6 c.p.m.) prepared as in *Protocol 1* or *Protocol 2*
- TE buffer (10 mM Tris–HCl, pH 8.0, 1 mM EDTA)
- Prehybridization buffer (100 ml):

deionized formamide[a]	50 ml
50 × Denhardt's solution	10 ml
20 × SSPE buffer	25 ml
10% SDS	2 ml
H_2O	to 100 ml final volume

    Prehybridization buffer without salmon sperm DNA may be prepared in advance and stored in aliquots at −20 °C

- Prehybridization buffer + salmon sperm DNA; to 20 ml of prehybridization buffer add 0.25 ml of freshly boiled salmon sperm DNA (boil 5 min immediately before adding to the prehybridization buffer); vortex to mix

Method

NB If several blots are being processed, treat each separately.

1. Wash the blot in prewash buffer for 1 h at 42 °C with gentle agitation.
2. Transfer the wet blot to a plastic bag (e.g. domestic freezer food bag)[b] and add 20 ml of prehybridization buffer + salmon sperm DNA that has been equilibrated to 42 °C.
3. Heat-seal the bag, excluding as many bubbles as possible, and prehybridize at 42 °C for at least 2 h.

Continued

Protocol 9. *Continued*

4. Start the hybridization, by cutting off a corner of the bag and removing the prehybridization solution into a small sterile flask. Adjust the volume to 5–10 ml.
5. Add the radiolabelled probe ($2-5 \times 10^6$ c.p.m.) to 0.1–0.2 ml of TE buffer in a microcentrifuge tube.
6. Heat the probe at 100 °C for 5 min and immediately add it to the prehybridization solution.
7. Mix it well and add the solution back to the plastic bag containing the blot.
8. Heat-seal the bag and hybridize the blot overnight with gentle agitation at an appropriate temperature[c].
9. Open the bag carefully and remove the radioactive solution. Dispose of it observing appropriate safety precautions.
10. Wash the blot for 30 min at room temperature in 1 litre of $1 \times$ SSC buffer containing 0.1% SDS.
11. Repeat step 10 two or three times.
12. Perform higher stringency washes, as necessary, by either repeating steps 10 and 11 at 42 °C or washing in $0.1 \times$ SSC buffer + 0.1% SDS at 68 °C.
13. Remove excess buffer from the blot but keep it damp and autoradiograph it at −70 °C using an intensifying screen and X-ray film (Kodak XAR5).

[a] Deionize the formamide as in *Protocol 3*, footnote a.
[b] Plastic hybridization cylinders (Hoefer) may also be used.
[c] Calculate the hybridization temperature according to *Protocol 3*, footnote c.

Synthetic oligonucleotide probes >18 nt in length can also be used in the Northern blot procedure. Oligonucleotides are routinely 5′ end-labelled (see *Protocol 2*). Oligonucleotide probes can be useful for detecting small internal sequence differences in RNA molecules that are otherwise identical.

4. Mapping of RNA branchpoints and 5′ ends of introns

4.1 Introduction

The chemical signature of group II intron splicing and nuclear pre-mRNA splicing is the RNA branchpoint structure (11) (see also Volume II, Chapter 7). At the centre of the structure resides a nucleoside, usually adenosine, which is linked via a 2′,5′ phosphodiester linkage to the 5′ nucleotide of the intron, as well as by the usual 3′,5′ phosphodiester bonds to its upstream and downstream sequences (12). RNA branchpoints accumulate in the characteristic RNA intermediates known as lariats of which two basic types exist, the RNA intermediate containing an intron-3′ exon structure and the final intron. Identification of RNA intermediates containing branchpoint(s) and mapping of the branchpoints are important procedures needed to characterize fully a novel RNA splicing mechanism.

i. Identification of lariat RNAs by denaturing polyacrylamide gel electrophoresis

The relationship between electrophoretic mobility and size of the RNA is markedly different for a lariat structure compared to a linear RNA of the same size. Hence, examination of electrophoretic mobilities on gels of different acrylamide concentrations will often enable lariat structures to be identified provisionally. As the pores of the gel decrease in size, the mobility of the lariat RNA decreases precipitously compared to RNA molecules that are linear (13,14). As a guide, electrophoresis in 4% and 7% polyacrylamide/7 M urea gels will show an abnormally abrupt change in migration of a lariat RNA with a loop size of approximately 400 nt and a linear length of 500–1200 nt. Higher percentage gels are recommended for smaller loop sizes. Chapter 1 (Section 2.3) provides protocols suitable for such an analysis.

ii. Standard procedure for mapping RNA branchpoints: primer extension analysis

Primer extension analysis is an analytical scale (100 000 c.p.m. of RNA substrate per reaction) procedure based upon the hybridization of a ^{32}P-labelled oligonucleotide primer downstream of the predicted branchpoint region, followed by primer extension towards the branchpoint. The chemical structure of the branchpoint nucleotide is an impediment to reverse transcriptase, resulting in chain termination one nucleotide downstream from the actual branchpoint. The high resolving power of this technique allows the precise mapping of branchpoints because primer extension products can be resolved on a sequencing gel adjacent to DNA sequencing ladders. An accuracy of ± one nucleotide can be expected. Best results are obtained when the primer is positioned approximately 30–60 nt downstream of the branchpoint. The primer extension procedure relies on control reactions containing debranching activity (see Section 4.2) in order to distinguish breaks in the RNA chain from genuine branchpoint structures.

4.2 Mapping RNA branchpoints by primer extension

The primer extension procedure can be applied successfully to a mixture of RNAs containing one or more lariat structures. Purification of individual lariat RNAs is not usually required. The primer extension analysis itself is identical to the standard procedure described in Section 3.5. It is followed by electrophoresis on a sequencing gel. However, as mentioned in Section 4.1, it is essential to include a debranching control to confirm the identity of branchpoint structures revealed by primer extension analysis.

The debranching extract is usually prepared from HeLa cells as described in *Protocol 10* (15,16). Prepare the extract fresh for each day's experiments. Following the debranching reaction, the sample can be stored for several weeks at $-20\,°C$ prior to primer extension analysis, although storage for only a few days or less is recommended. Primer extension reactions are then performed on both the debranched and untreated samples as described in *Protocol 7*.

Protocol 10. Preparation of debranching extract[a]

Equipment and reagents

- HeLa spinner culture cells grown in Joklik's minimal essential medium + 5% horse serum
- PBS (0.13 M NaCl, 20 mM potassium phosphate, pH 7.4)
- Dounce pattern glass homogenizer with a tight-fitting pestle (Wheaton)
- Dialysis buffer (20% glycerol, 0.1 M KCl, 20 mM Hepes–KOH, pH 7.6, 0.2 mM EDTA)
- 0.1 M EGTA
- 50 mM $CaCl_2$
- Micrococcal nuclease (Sigma)
- Buffer A (10 mM Hepes–KOH, pH 7.6, 1.5 mM $MgCl_2$, 10 mM KCl, 0.5 mM DTT)
- Buffer B (0.3 M Hepes–KOH, pH 7.6, 1.4 M KCl, 30 mM $MgCl_2$)
- Sorvall centrifuge and SS34 rotor, Beckman ultracentrifuge and SW41 rotor plus tubes (or equivalents)

Method

1. Grow HeLa cells to a density of 3×10^5 cells/ml in Joklik's MEM and 5% horse serum, and measure the cell density using a haemocytometer.
2. Using 1 litre centrifuge bottles, pellet the cells at $1000\,g$ for 10 min at 4 °C and discard the supernatant. *Beware*, the cell pellet is very loose. Keep the cells on ice.
3. Consolidate the cells into one graduated 50 ml centrifuge tube by resuspending in ice-cold PBS. Pellet the cells as in step 2. Note the packed cell volume.
4. Add 5 packed cell vol. of ice-cold buffer A and resuspend the cells gently.
5. Stand the cells on ice for 10 min.
6. Pellet the cells at $1500\,g$, for 10 min at 4 °C. Discard the supernatant.
7. Add 2 packed cell vol. of buffer A and resuspend the cells gently on ice.
8. Transfer the cell suspension to a chilled glass homogenizer. Homogenize the cells with a tight-fitting pestle until $\geq 90\%$ of the cells are lysed. Keep the mixture on ice. To check for cell lysis and release of nuclei, remove a small volume of the mixture into a fresh tube and add 9 vol. of PBS + 0.125% trypan blue. The nuclei will stain with trypan blue whereas intact cells will not.
9. Transfer the suspension of nuclei into a chilled 40 ml centrifuge tube and spin at $2000\,g$ for 10 min at 4 °C.

Continued

2: Characterization of RNA

Protocol 10. *Continued*

10. Transfer the supernatant, using a 10 ml plastic pipette, into a 50 ml tube and store it on ice.
11. Centrifuge the nuclear pellet in the same tube for 20 min at 4 °C at 16 000 r.p.m. in a Sorvall SS34 rotor to condense the nuclear pellet.
12. Combine the second supernatant with that from step 10. At this step the nuclear pellet can be used to prepare Dignam nuclear extract (16), or discarded.
13. Measure the volume of the combined supernatants and add 0.11 vol. of buffer B. Transfer the mixture into chilled polypropylene tubes for the SW41 rotor.
14. Centrifuge the samples at 30 000 r.p.m. at 4 °C in the SW41 rotor for 1 h (100 000 g).
15. Dialyse the supernatant in 1 litre of dialysis buffer for 2.5 h at 4 °C.
16. Repeat step 15. The resulting solution is the S100 extract.
17. Store the S100 extract in small volumes (0.2 ml) at −70 °C after flash-freezing the extract in liquid nitrogen.
18. Prepare the debranching extract by adding the following to a 1.5 ml microcentrifuge tube:
    - S100 extract   200 µl
    - 50 mM $CaCl_2$ 4 µl
    - micrococcal nuclease   4 µl
19. Incubate the mixture at 30 °C for 30 min.
20. Stop the reaction by transferring the tube to an ice bucket and add 8 µl of 0.1 M EGTA, pH 8.0[b].
21. Keep the debranching extract on ice until required.

[a] Based on refs 15 and 16.
[b] A molar excess of EGTA over $CaCl_2$ is needed to ensure that the micrococcal nuclease remains inactive during the subsequent debranching reaction (see *Protocol 11*).

Having prepared the debranching extract, carry out the debranching reaction itself as described in *Protocol 11*. The amount of RNA substrate required for branchpoint determination is 25–250 ng, the quantity usually present in a typical splicing reaction (see Chapter 3, Section 3.2). However, this assumes that a reasonable efficiency of splicing is achieved. The time course of the splicing reaction should be investigated to determine at what point in the reaction the lariat species are most enriched. In addition, the splicing reactions should be scaled up or down according to the amount of material needed for branchpoint determination.

> **Protocol 11.** Debranching reaction
>
> *Reagents*
>
> - Sample containing RNA splicing intermediates for analysis (see Chapter 3, Section 3.2)
> - Proteinase K (Boehringer Mannheim)
> - 2 × proteinase K buffer (0.2 M Tris–HCl, pH 7.6, 25 mM EDTA, 0.3 M NaCl, 2% SDS; containing 5 mg/ml proteinase K)
> - Debranching extract, freshly prepared as in *Protocol 10*
> - Debranching buffer (0.1 M KCl, 20 mM Hepes–KOH, pH 7.6, 8 mM EDTA, 0.5 mM DTT, 20% glycerol)
> - HHS buffer, phenol, chloroform, and reagents for ethanol precipitation of RNA as in *Protocol 3*.
>
> *Method*
>
> 1. Terminate the splicing reactions by adding an equal vol. of 2 × proteinase K buffer containing 5 mg/ml proteinase K followed by incubation at 30 °C for 10 min.
> 2. Add 400 µl of HHS buffer. Extract with phenol and with chloroform followed by ethanol precipitation (*Protocol 3*, steps 8–10).
> 3. Resuspend the dried RNA pellet in ≥ 2 µl of sterile H_2O.
> 4. Combine 2 µl of RNA from step 3 and 25 µl of debranching mixture in a microcentrifuge tube, for each debranching reaction.
> 5. Incubate at 30 °C for 30 min.
> 6. Transfer the reaction tubes to an ice bucket and adjust the volume to 400 µl with ice-cold HHS buffer.
> 7. Extract and recover the RNA from each reaction as described in *Protocol 3*, steps 8–10.

Figure 3 shows an example of the results which can be expected from primer extension analysis of branch points. Primer extension analysis reveals three extension products (I, II, and III). Products II and III disappear upon treatment with debranching extract (compare lanes 180' and DB) indicating that these products are generated from RNAs containing true branchpoints. Product I, however, does not disappear under debranching conditions indicating that it corresponds to a linear RNA molecule. The branchpoint in product III can be mapped by comparison to the DNA sequencing ladder (lanes U, G, C, A). The primer extension stops at an adenosine nucleotide, one residue downstream of the actual branchpoint position, which is also an adenosine residue (see lane A). Quantitative comparison of II and III indicates that the former results from a minor RNA intermediate.

4.3 Mapping the 5' end of an intron

The primer extension procedure can also be used to map the 5' end of an intron. This is accomplished by hybridizing a primer to the intron, within the loop

2: Characterization of RNA

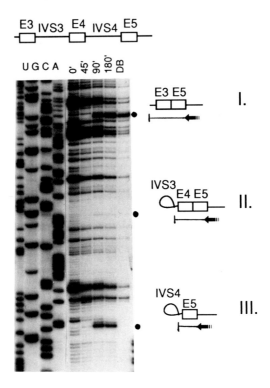

Figure 3. Branchpoint determination by primer extension. A splicing substrate, shown schematically at the top, was incubated under *in vitro* splicing conditions (see Chapter 3, Section 3.2) for times as indicated, lanes 0', 45', 90', and 180'. A duplicate 180' reaction sample was also treated under debranching conditions (see *Protocol 11*) prior to the primer extension reaction, (lane DB). Dideoxy sequencing ladders are shown at left (lanes U, G, C, and A) for the corresponding plasmid DNA. Primer extension products are indicated by the filled circles at the right. All samples were primer extended with a ^{32}P-labelled primer which is complementary to IVS5 sequences located downstream of exon 5. Schematic structures at the right indicate the progression of the primer extension reaction which begins at the primer (arrow) and ends at the branchpoint or 5'-end of the RNA (vertical line). The primer is hybridized to the RNA template prior to the addition of reverse transcriptase. Product I corresponds to the linear spliced product RNA in which exon 3 is joined to exon 5 (lanes 90', 180', and DB). Product II, which is present at low abundance, corresponds to the intermediate in which IVS3, in a lariat form, is joined to exons 4 and 5 (lanes 90' and 180'). Product III corresponds to the intermediate in which IVS4, in a lariat form, is joined to exon 5 (lanes 90' and 180'). In the case of Product I, the primer extension stops at the 5'-end of the RNA substrate, whereas for products II and III the primer extension stops at the branchpoint. Additional primer extension products present in these reactions are resolved at different positions on the same sequencing gel and are not shown. Background primer extension products are due to spurious RNA cleavages that are unrelated to splicing.

portion of the lariat, followed by primer extension to the 5' end of the intron. If the lariat form of the intron is used as the template for the reaction, the primer extension will stop one nucleotide downstream of the 5' end, since this end is attached to the branchpoint structure. Alternatively, the lariat intron can first be debranched and the resulting linear form used in the primer extension reaction to map the 5' end. The protocols necessary for mapping the 5' end of the intron are those described in Sections 3.5 and 4.2.

Acknowledgements

The author is grateful to former and current colleagues for their contributions to the methods described in this chapter.

References

1. Berk, A. J. and Sharp, P. A. (1977). *Cell*, **12**, 721.
2. Vogt, V. M. (1973). *Eur. J. Biochem.*, **33**, 192.
3. Sambrook, J., Fritsch, E. F., and Maniatis, T. (ed.) (1989). In *Molecular cloning: a laboratory manual*, (2nd edn). Cold Spring Harbor Press, Cold Spring Harbor, NY.
4. Bolton, E. T. and McCarthy, B. J. (1962). *Proc. Natl Acad. Sci. USA*, **48**, 1390.
5. Krämer, A., Keller, W., Appel, B., and Lührmann, R. (1984). *Cell*, **38**, 299.
6. Chirgwin, J. M., Przybyla, A. E., MacDonald, R. J., and Rutter, W. J. (1979). *Biochemistry*, **18**, 5294.
7. Mullis, K. and Faloona, F. A. (1987). In *Methods in enzymology*, Vol. 155, (ed. R. Wu), p. 335. Academic Press, San Diego.
8. Saiki, R. K., Gelfand, D. H., Stoffel, S., Scharf, S. J., Higuchi, R., Horn, G. T., Mullis, K. B., and Erlich, H. A. (1988). *Science*, **239**, 487.
9. Innis, M. A., Gelfand, D. H., Sninsky, J. J., and White, T. J. (ed.) (1990). *PCR protocols. A guide to methods and applications*. Academic Press, New York.
10. Nasim, F. H., Spears, P. A., Hoffmann, H. M., Kuo, H.-C., and Grabowski, P. J. (1990). *Genes Dev.*, **4**, 1172.
11. Wallace, J. C. and Edmonds, M. (1983). *Proc. Natl Acad. Sci. USA*, **80**, 950.
12. Konarska, M. M., Grabowski, P. J., Padgett, R. A., and Sharp, P. A. (1985). *Nature*, **313**, 552.
13. Grabowski, P. J., Padgett, R. A., and Sharp, P. A. (1984). *Cell*, **37**, 415.
14. Ruskin, B., Krainer, A. R., Maniatis, T., and Green, M. R. (1984). *Cell*, **41**, 833.
15. Ruskin, B. and Green, M. R. (1985). *Science*, **229**, 135.
16. Dignam, J. D., Lebovitz, R. M., and Roeder, R. G. (1983). *Nucleic Acids Res.*, **11**, 1475.

3

Splicing of mRNA precursors in mammalian cells

IAN C. EPERON and ADRIAN R. KRAINER

1. Introduction

Splicing has attracted and intrigued researchers ever since the first reports of split genes. The first studies attempted to define the signals which were required for splicing. This work also produced some indications of the importance of introns and splicing for efficient expression. Later experiments on *cis*-acting signals, particularly in mammalian systems, have been directed towards understanding the mechanisms by which the correct splice sites can be selected in the face of a plethora of candidates. In some cases, of course, the process of selection is affected by the stage of the cell cycle or cell differentiation. Most of this work has been done *in vivo*, involving the introduction of mutant substrates into cells and subsequent analysis of the mRNA expressed.

More recently, two major developments have provided the means to analyse the molecular processes involved in splicing. The first of these is splicing *in vitro*, using extracts from nuclei or whole cells. This technique led to the discovery that splicing appeared to proceed in two steps, the first of which produced intermediates comprising the 5' exon and a lariat form of the intron–3' exon moiety. Furthermore, splicing *in vitro* provided opportunities to test existing hypotheses about the roles of small nuclear ribonucleoprotein particles (snRNPs), to study their incorporation with the substrate into multicomponent assemblages, and to fractionate the extracts in order to isolate splicing factors.

The second key development has been the use of genetic analysis in yeast in combination with splicing *in vitro*. This technique has been critical for uncovering a number of splicing components in addition to the snRNPs, and it has enabled the assembly and interactions of snRNPs to be dissected. More recently, it has been used to study the factors involved in splice site recognition.

This chapter describes some of the techniques that are fundamental to studying pre-mRNA splicing. Section 2 describes the introduction of mutant genes into mammalian cells and methods for analysing the mRNA produced. The preparation of splicing extracts from mammalian cells is presented in Section 3, together with a brief discussion of methods for depleting the extracts of specific components. The fractionation of extracts and the identification and purification of components are described in Section 4.

2. Analysis of pre-mRNA splicing in mammalian cells

2.1 Expression of exogenous genes
In any investigation, a choice has to be made of the cell type, vectors, transfection method, and splicing construction (gene organization).

2.1.1 Choice of cell line
Most investigations have studied basic splicing processes in commonly-used established cell lines which are often presumed to be representative of non-specialized cells. The lines used most often are primate lines which can grow in monolayers and are robust, such as HeLa, 293, and COS-1 cells. These cells can be transfected with a high efficiency, which allows analysis after transient expression rather than after selection of stable transfectants. For specialized studies, where the splicing reaction is specific to the tissue or stage of development, an appropriate tissue-specific cell line or transgenic organism would be a prerequisite.

2.1.2 Vectors
An enormous range of vectors is available for transient expression studies. Many of these are listed in ref. 1, and others are available from a number of commercial companies. To ensure expression of a cloned gene fragment, it should be inserted between a strong enhancer–promoter and a polyadenylation site. The enhancer and promoter must be appropriate for the cell line used. In HeLa cells and COS-1 cells, an SV40 enhancer–promoter can produce high levels of expression. Other very effective enhancer–promoters are derived from human cytomegalovirus and the β-actin gene. A number of other enhancers are used for expression in specific tissues or cell lines.

Transient expression is often enhanced if the vector carries an SV40 origin of replication. This will be effective only in the presence of the SV40 large T-gene product, which can be provided either from a gene on the same plasmid or from a co-transfected plasmid. COS cells are particularly convenient for use with vectors carrying an SV40 origin because the cells express the large T antigen from an integrated gene (2). There have been some reports of inappropriate splicing when genes exhibiting alternative splicing patterns are expressed to high levels (3, 4). In such cases, it may be preferable to create cell lines such that every cell expresses the test gene but at a relatively low level.

In addition to the essential plasmid origin of replication and a selectable marker for growth in *Escherichia coli*, it may be useful if the vector includes an M13 or fd origin, for production of single-stranded DNA, and SP6 or T7 RNA polymerase promoters. Except in cases where one of the alternative splicing patterns being studied involves retention of an intron in the mRNA, it is unlikely that it will be essential for the vector to contain introns within the test transcription unit. Indeed, the presence of a short intron in the vector might lead to the activation of cryptic splice sites in the test fragment (5).

2.1.3 Transfection of cells

The choice of the transfection method depends upon the cells used and the laboratory facilities available. Methods used for transient expression involve mediating DNA uptake by calcium phosphate, DEAE-dextran or cationic lipids, or the use of electroporation. Calcium phosphate-mediated uptake is the most common method used for HeLa cells, and is described in *Protocol 1*.

Protocol 1. Calcium phosphate–DNA cotransfection of adherent HeLa cells

Reagents
- Growth medium: Dulbecco's modified Eagle's medium (DMEM) supplemented with 2 g/l NaHCO$_3$, 2.5% newborn calf serum and 2.5% foetal calf serum
- CaCl$_2$ solution: mix 100 ml of 1 mM Tris–HCl, pH 7.5, 0.1 mM EDTA, and 15 ml of 2 M CaCl$_2$
- Recombinant plasmid DNA (0.1–0.33 mg/ml) incorporating the DNA sequence to be studied; this should be free of *E. coli* chromosomal DNA or RNA
- 2 × HBS: dissolve 4 g NaCl, 0.185 g KCl, 50 mg Na$_2$HPO$_4$ (anhydrous), 0.5 g glucose and 2.5 g Hepes in 200 ml of distilled water; adjust the pH to 7.05 with NaOH, and make the volume up to 250 ml with water; sterilize the solution by passing it through a 0.2 μm membrane filter
- DMSO–DMEM solution (25% DMSO and 75% DMEM, without serum); prepare this fresh, and warm it to 37 °C just before use.

Method

1. Replate the HeLa cells (in 90 mm diameter dishes) 24 h before use, such that they will be about 50% confluent when transfected.
2. Replace the medium with fresh growth medium 4 h before transfection.
3. For each dish, mix 600 μl CaCl$_2$ solution and 60 μl of recombinant plasmid DNA solution in a sterile 1.5 ml microcentrifuge tube.
4. Add 600 μl of 2 × HBS to the CaCl$_2$–DNA mixture. Mix and leave the tube on ice for 10 min. A fine precipitate should form.
5. Take each dish of cells out of the incubator and add the DNA mixture in drops distributed across the surface of the dish. Shake the dish gently and return it to the incubator.
6. Replace the medium after 16–24 h with pre-warmed DMSO–DMEM solution. Leave the dish for 3 min.
7. Remove the DMSO–DMEM solution by aspiration and wash the monolayer twice with growth medium.
8. Add fresh growth medium and replace the dish in the incubator.
9. Harvest the cells after another 24–48 h (Chapter 1, *Protocol 9*).

The DEAE-dextran method is usually used for COS-1 cells, but the CaCl$_2$-DNA procedure (*Protocol 1*) also gives reliable results with COS-1 cells when it is modified as follows to reduce cell mortality:

- use a DMSO-DMEM solution consisting of 15% DMSO, 85% DMEM
- after adding the DNA, incubate the cells for only 5 h before adding DMSO

If the DMSO step is still toxic despite these measures, 20% glycerol in DMEM can be used instead of DMSO-DMEM. Other methods for transfection, and more details about the variables to be considered, can be found elsewhere (for example, refs 6 and 7).

2.1.4 Gene organization

Where alternative patterns of splicing are being investigated, it is usual to simplify the experimental manipulations and subsequent analysis by expressing only the affected introns and exons within the vector's test transcription unit. Despite the complexity often seen in mRNA splicing, this is a secure strategy: sequences beyond the constitutive exons flanking the region in question do not seem to determine the local patterns of alternative splicing. Thus, most strategies for identification of the signals that determine splicing patterns have involved insertion of a gene fragment bounded by its constitutive exons into the exon sequences of the vector. It should be noted, however, that the flanking exon sequences might affect the choice of alternative exons (8) and that, in at least one case, a splice site at the extremity of a flanking constitutive exon was important (9). In the latter case, an exon fusion strategy would not have revealed this, and it may be better to include intron sequences adjacent to the constitutive exons. For this reason, and for the rapid elimination of possible splicing signals, it may be desirable to make intron fusions, inserting exons with flanking intron portions inside a vector intron. Thus, a vector with unique cloning sites in exon and intron sequences may be helpful.

Finally, the presence of transcripts similar to the test transcript in the cell line chosen should be considered. If this is likely to confuse attempts to map the transcripts derived from transfected DNA, ensure that it is feasible to use flanking vector sequences for RNA mapping or introduce unique 'tags' into your test gene fragment (Section 2.2.2).

2.1.5 Transgenic animals

Where there is tissue-specific regulation of alternative splicing, it is sometimes possible to view one splicing pattern as a non-specific or default pattern and the other as being more restricted and therefore tissue-specific. In some cases, the gene is expressed in only a small number of tissues, and it is not possible to determine which tissue-specific pattern is regulated and which is (or might be) an unregulated default pattern. An appropriate strategy here is to make transgenic mice which express the gene in all tissues. This is exemplified by work on the calcitonin gene, which expresses calcitonin mRNA in thyroid C cells and mRNA for a calcitonin gene-related peptide in neurones: in most tissues of the transgenic mouse, there was a preponderance of calcitonin mRNA (10). A detailed description of transgenic animal technology is beyond the scope of this chapter. Strategies and protocols will be found in another volume of this Series (11).

Establishing a default pattern allows a simplifying heuristic assumption that the specific splice is either incompetent in default cells or it is in competition with a stronger default splice. It might be predicted that the specific tissue would, in these respective cases, need either to activate the specific splice or to repress the default splice. The two mechanisms can be distinguished in a formal sense with simple systems in which one constitutive exon splices to either of two alternative sites. If the specific site was, in principle, in competition with the default splice site then deletion of the default site would be expected to allow use of the specific site in default cells. This indicates that the specific site would not need activation in the specific tissue. This logic has been described more fully elsewhere (12, 13).

2.2 Analysis of RNA expressed

2.2.1 Isolation of RNA from transfected mammalian cells

The basic procedures used are described in Chapter 1. For nuclease S1 analysis it is not necessary to select poly(A)$^+$ RNA, but it has been found in this laboratory that the use of poly(A)$^+$ RNA can give clearer results than total RNA in reverse transcriptase–PCR reactions (see Section 2.2.3). To purify poly(A)$^+$ RNA from a large number of samples, it is particularly advantageous to use Hybond™-mAP messenger affinity paper (Amersham), as described in Chapter 1 (Section 3.3).

2.2.2 Analysis of RNA by nuclease S1

The method used for mapping RNA depends on the abundance of the mRNA and the pattern of splicing being studied. When the RNA is fairly abundant, mapping with nuclease S1 is preferred, because of the ease with which quantitative data can be obtained. The strategy is outlined in *Figure 1*. The use of end-labelled genomic probes provides reliable, quantitative measurements of the relative levels of alternative splice sites if the alternative sequence is contiguous in the genomic sequence with the site of labelling (see *Figure 1*). However, such probes cannot be used for the detection of discrete alternative exons. Probes consisting of cDNA made from the expected RNA isoforms are useful in these circumstances, but there are two points to watch:

(a) Each probe can provide quantitative evidence only for the relative proportions of products it detects; intensities cannot be compared between tracks when two alternative patterns of skipping (i.e. mutually exclusive splicing) are being detected with the two possible cDNA isoforms.

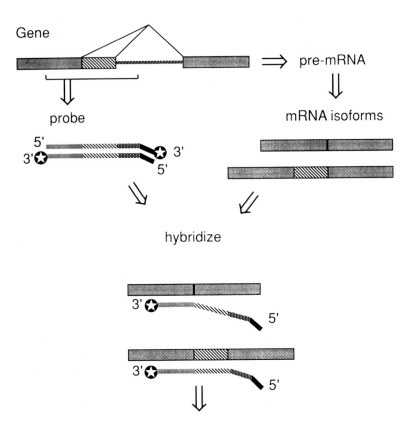

Figure 1. Mapping exon boundaries with probes extending into the intron. The diagram shows the use of 3′ end-labelled probe to detect alternative 5′ splice sites. A 5′ end-labelled probe would be used to detect 3′ splice sites. Universally-labelled probes can be used to detect both 5′ and 3′ splice sites, if allowances are made during quantitative analysis for the number of labelled nucleotides in each fragment protected. Exon sequences in the gene and in the mRNA are shown boxed; the non-constitutive portion of exon is striped and the intron is shown as a thinner line. The double-stranded DNA probe is drawn in the same patterns. The additional 'tail' in the probe (solid line) is not complementary to the RNA; it is introduced during PCR or is derived from vector sequences in order to distinguish fully-protected products (where the intron has not been removed from the RNA) from undigested probe. The asterisk indicates the site of labelling.

(b) Extra sequences in the RNA can loop out, leaving a fully protected probe (14) as shown in *Figure 2A*. The RNA containing the extra sequences would be undetected or its abundance underestimated. The authors find that this is a problem only where the extra sequence is short, as is usually the case when an extra exon has been incorporated rather than when an intron has been retained. Thus, to avoid this problem, a cDNA probe should be prepared from the longest isoform where possible.

For the above reasons, it is useful, and often imperative, to use an independent method (Section 2.2.3) for mapping the transcripts.

End-labelled probes can be prepared from restriction fragments or PCR products. In both cases, either the 5' end or the 3' end can be labelled. The products of a PCR have either a blunt end or a protruding, template-independent, 3' nucleotide (15). Thus, for labelling the 3' end of a PCR product with a DNA polymerase, it is necessary to create a recess, either by restriction enzyme cleavage or controlled digestion with T4 DNA polymerase (in the presence of one dNTP). Note that the labelled end of such probes must be complementary to exon sequences and, if exogeneous transcripts are to be mapped without interference from endogeneous transcripts, such sequences must be unique to the exogeneous gene. This can be arranged either by incorporating new sequences (such as linkers) in the exogeneous cloned gene or by including flanking vector sequences in the probe. It can be helpful if the last ~ 30 nt at the other end of the probe is not complementary to the RNA. This permits the fully-protected probe to be distinguished from undigested probe, allowing incomplete digestion by nuclease S1 to be diagnosed and the undigested probe to be ignored in any quantitative analysis. The use of nuclease S1 is included as *Protocol 4*; a similar procedure is also given in Chapter 2 (Section 3.2), where more details are given of difficulties that may be encountered.

Labelling the 3' end of a DNA restriction fragment using an $[\alpha\text{-}^{32}P]$ deoxyribonucleotide triphosphate and DNA polymerase I (Klenow fragment) is described in Chapter 2 (*Protocol 1*). The preparation of a 5' end-labelled probe, appropriate when the 5' end of the probe will lie within an exon, using PCR technology is described below in *Protocol 2*. This describes typical PCR conditions suitable for 20mer oligonucleotide primers and a product of about 300 bp. However, the best annealing condition, $MgCl_2$ concentration and primers may need to be checked in pilot experiments. The oligonucleotide primer complementary to the mature RNA is labelled; the other primer should be complementary at its 3' end for 15–20 nt to the genomic or cDNA PCR template, and other sequences may be introduced at the 5' end of this primer to provide a probe sequence that does not match the RNA (see above).

3: Splicing mRNA precursors

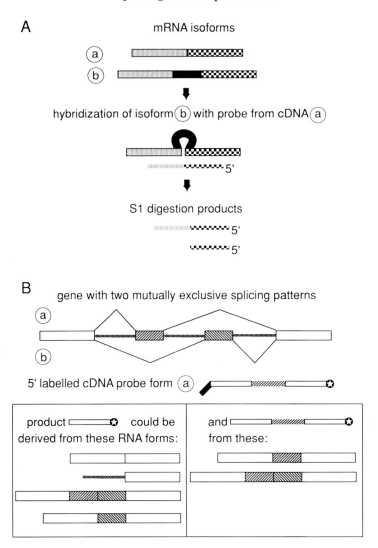

Figure 2. Some sources of uncertain assignments when using cDNA sequences as nuclease S1 probes. (A) Use of a probe which does not include additional internal sequences found in some RNA isoforms allows hybrids to form in which the extra RNA sequences are looped out, but the probe remains almost completely double stranded and is not digested efficiently by nuclease S1 opposite the site of insertion of the extra RNA sequences. The constitutive exons in the two isoforms of mRNA, **a** and **b**, are stippled and chequered; the optional exon is black. The complementary strand of the DNA probe derived from isoform **a** is shown. The site of labelling is not shown. (B) Ambiguous assignments from nuclease S1 analysis with cDNA probes of alternative patterns of exon incorporation. In the example shown here, two constitutive exons (white boxes) flank two alternative exons (striped boxes), whose incorportion is mutually exclusive (the two patterns of splicing **a** and **b** are shown connecting the exons incorporated). The complementary strand of the DNA probe derived from isoform **a**

Protocol 2. Preparing a 5' end-labelled probe by PCR

Equipment and reagents

- Oligonucleotide primers (>2 pmol/µl)
- [γ-^{32}P]ATP (~3000 Ci/mmol, ~10 mCi/ml)
- T4 polynucleotide kinase (10 units/µl)
- 10 × C buffer (0.5 M Tris–HCl, pH 7.5, 0.1 M MgCl$_2$, 10 mM DTT)
- *Taq* polymerase (>1 unit/µl)
- 10 × PCR buffer (0.5 M KCl, 0.1 M Tris–HCl, pH 7.5, 15 mM MgCl$_2$, 2 mM dATP, 2 mM dCTP, 2 mM dGTP, 2 mM dTTP, 2 mg/ml BSA). Check that this is suitable for the enzyme being used
- Template plasmid (1–100 ng in 0.1–10 µl). See Chapter 1 (*Protocol 1*) for preparation
- PCR thermocycler apparatus
- Materials for denaturing polyacrylamide gel electrophoresis (Chapter 1, *Protocol 8*)
- Elution buffer (0.5 M sodium acetate, pH 5–6, 1 mM EDTA, 0.2% SDS)

Method

1. In a 0.5 ml microcentrifuge tube, mix 10 pmol of oligonucleotide complementary to the mRNA with 30 µCi of [γ-^{32}P]ATP (~10 pmol) and 1 µl 10 × C buffer.
2. Add 5 units of T4 polynucleotide kinase and incubate the mixture at 37 °C for 30–60 min.
3. Heat the mixture at 70 °C for 10 min to inactivate the enzyme.
4. Prepare a 25 µl PCR reaction in a 0.5 ml microcentrifuge tube by mixing 2.5 µl of 10 × PCR buffer, 10 pmol of the RNA-sense primer, 0.5 units of *Taq* polymerase, 1–100 ng of template plasmid, and water to 25 µl final volume.
5. Add the kinase reaction mixture (step 3) to the PCR mixture (step 4); the final volume will be ~35 µl. Add 4 drops of paraffin oil.
6. Amplify the sequences for 10–30 cycles; 10 cycles should amplify the target sequence 100-fold to 1000-fold, depending on the efficiency of the PCR reaction. Standard conditions for each cycle with 20mer primers and a target of 300 bp would be 94 °C, 0.5 min; 50 °C, 0.5 min; 72 °C, 1 min. The final extension is usually 10 min.

Continued

---

is shown; the asterisk shows the site of labelling. After hybridization and digestion by nuclease S1, splicing pattern **a** should be revealed by production of the probe fragment shown at the top of the right-hand box. However, for the reasons shown in panel A, the fully-protected probe could be produced even if extra sequences were incorporated by perturbation of the splicing pattern. The right-hand box shows the expected mRNA isoform **a** and one aberrant mRNA which could result in the same probe fragment. Likewise, the shorter probe product shown in the left-hand box would arise from hybridization to mRNA isoform **b** (bottom of the left-hand box). This is because the central portion of the probe would not be complementary to the alternative exon incorporated, and thus this part of the probe would be digested. However, a number of other mRNA isoforms (also shown in the left-hand box) might have arisen which also be consistent with this pattern.

3: Splicing mRNA precursors

Protocol 2. Continued

7. Remove the paraffin oil by aspiration and extract any remaining oil by vortexing the reaction mixture with a larger volume of ether.
8. Purify the labelled probe by following the procedure described in Chapter 1 (*Protocol 8*) in which the nucleic acids are precipitated with ethanol and separated by denaturing polyacrylamide gel electrophoresis. The labelled band is located by autoradiography. Excise the probe band and soak it in 0.25 ml of elution buffer either at 55 °C for 45 min or at 4 °C overnight.
9. Add 2.5 vol. of ethanol at ambient temperature to the eluate. Do not chill. This can be used directly for hybridization to the RNA.

Protocol 3 describes nuclease S1 analysis using end-labelled probes prepared as described above. Further details and problem-solving advice are given in Chapter 2 (Section 3.2).

Protocol 3. Hybridization and digestion with nuclease S1

Reagents

- End-labelled probe prepared as described in *Protocol 2* or Chapter 2 (*Protocol 2*). The probes should be kept in 70% ethanol at ambient temperature.
- Test RNA (1 ml of total RNA from cells in a 90 mm dish)[a]
- Positive control RNA (200 μl containing 0.1 μg RNA synthesized *in vitro*, or residual test RNA from previous experiments in which the RNA had been detected successfully with the same probe)[a]
- Negative control RNA (200 μl of RNA isolated from untransfected cells, or 5 μg tRNA)[a]
- Hybridization buffer (80% deionized formamide[b], 0.4 M NaCl, 40 mM Pipes, pH 6.4, 1 mM EDTA)
- Nuclease S1 digestion buffer (0.28 M NaCl, 50 mM sodium acetate, pH 4.6, 4.5 mM $ZnSO_4$)
- Nuclease S1 (Pharmacia)
- Single-stranded DNA (commercial salmon sperm DNA at 10 mg/ml in water, which has been incubated in boiling water for 20 min and frozen at −20 °C in aliquots)
- Carrier RNA (rRNA at 10 mg/ml)
- Materials for denaturing polyacrylamide gel electrophoresis and autoradiography (Chapter 1, *Protocol 8*)

Method

1. Dispense into four 1.5 ml microcentrifuge tubes:
   (i) 200 μl probe + 200 μl test RNA[c]
   (ii) 200 μl probe + 200 μl positive control RNA
   (iii) 200 μl probe + 200 μl negative control RNA
   (iv) 20 μl probe (set aside at −20 °C until step 7)

Continued

Protocol 3. *Continued*

2. Centrifuge samples (i) to (iii) for 15 min in a microcentrifuge, and wash the pellets in ethanol.
3. Dissolve each pellet in 30 µl of hybridization buffer.
4. Heat the tubes at 80 °C for 10 min. Immediately transfer them to a hybridization bath. The optimal hybridization temperature may need to be determined in a pilot experiment. Start at 50 °C and increase the temperature in steps of 2 °C. At the optimal temperature, DNA:RNA hybridization will be favoured over DNA:DNA renaturation. Hybridize the samples for 2–20 h.
5. Add 20 µg of single-stranded carrier DNA and 1000 units (usually about 20 µl) of nuclease S1 to 1 ml of nuclease S1 digestion buffer. Keep this mixture on ice.
6. Open the tubes, at the end of hybridization (step 4), in the hybridization bath. Add 0.25 ml of nuclease S1 mixture to the first tube in the bath, remove it, vortex it, and immediately place it in a 37 °C bath. Repeat this with the other tubes in turn.
7. Remove all the tubes after 20–30 min. Add 10 µg of carrier RNA and 0.7 ml ethanol to each tube and mix. Centrifuge these tubes, and the sample of probe set aside in step 1 (iv), as in step 2 to pellet the nucleic acids (cooling is unnecessary).
8. Wash the pellets with absolute ethanol and centrifuge again. Air dry the pellets.
9. Analyse the samples and controls by denaturing polyacrylamide gel electrophoresis (Chapter 1, *Protocol 8*, Chapter 4, *Protocol 11*). For best results, the gel should comprise 20% formamide (v/v) as well as 7 M urea. Radioactive length markers should also be loaded.

[a] The volumes given are for RNA in 70% ethanol (v/v), 83 mM sodium acetate. The samples should be brought to ambient temperature before use.
[b] Deionize the formamide using a mixed-bed ion-exchange resin and store it in aliquots at −20 °C.
[c] The probe should be in a substantial molar excess with respect to the test mRNA.

Nuclease S1 analysis may also be done using single-stranded DNA probes. These can be prepared in a variety of ways, the commonest of which involves priming on a single-stranded, RNA-sense M13 template. Another method involves linear amplification with *Taq* polymerase and a single primer from a template of defined length. As with the PCR reactions described above, a labelled primer can be used or labelled nucleotide can be incorporated. However, these probes can produce a rather higher background in nuclease S1 analysis, even though the hybridization temperature is less critical. RNA probes prepared by transcription *in vitro* can also be used for mapping RNA. The basic principles for the use of these probes are similar to those of the nuclease S1 procedure, but selective digestion of single-stranded RNA is achieved by a mixture of RNases (A and T1) rather than nuclease S1 (16).

2.2.3 Analysis of RNA by reverse transcription–polymerase chain reaction

Mapping RNA by reverse transcription followed by PCR (RT–PCR) has a number of applications in the analysis of mRNA isoforms. Reverse transcription alone is of limited value with complicated genes because the distance from the 5' end of the transcript is usually so long that either the efficiency of the reaction or the electrophoretic resolution is poor. However, if the reverse transcription reaction is followed by a PCR between flanking constitutive exons then the signal will be amplified and the products will be shorter. The individual bands may be excised from a gel and sequenced. Alternatively, the products can be cloned and sequenced. Such a method is particularly useful for quantitative analysis when one pair of primers is capable of detecting all the isoforms of interest. This is very important for:

(a) Analysis of mRNA where alternative splicing could produce an unknown number of combinations of exons between flanking constitutive exons. Sequencing the PCR products would define the available exons.
(b) Analysis of alternative splicing from exogeneous or endogeneous genes where mapping by nuclease S1 is ambiguous because one band could be derived either from mRNA incorporating an alternative exon or from unspliced RNA. A similar ambiguity might arise in mutually exclusive splicing if a mutation caused incorporation of both alternative exons; nuclease S1 mapping with cDNA probes of both expected isoforms might produce fully-protected bands due to the looping-out of extra RNA sequences (*Figure 2B*; ref. 14). The use of RNase H and nuclease S1 analysis (see Chapter 2, Section 3.3) might not resolve these problems unless probes are prepared which contain all the possible extra sequences.
(c) Mapping scarce RNA from, for example, specific cell lines which can be transfected only with low efficiency.

Quantitative analysis of isoforms using PCR has to be approached with caution. For each isoform, an equation describing the early (exponential) phase of amplification is:

$$N_n = N_o \cdot eff^n \qquad [1]$$

where N_o refers to the number of molecules of target DNA present at the start of the reaction, N_n is the yield (the number of molecules after n cycles) and *eff* refers to the efficiency of the reaction (the ratio of the number of molecules present after each cycle compared with the number at the beginning, the value of which is expected to be between 1 and 2).

A plot of log N_n against n should give a straight line with slope = log*eff*. At later stages, the amplification efficiency declines progressively as components become limiting.

If, for two isoforms *a* and *b*, the values of *eff* are the same then, at any cycle within the exponential phase of the PCR, the ratio of yields of *a* and *b* will be the same as the ratio of the initial cDNA isoforms. It is usual to measure the ratios and the absolute yields at several successive cycles to ensure that the reaction is in this phase. However, if the efficiencies differ at all, then the relative yields of two products may change from cycle to cycle because:

$$\log(N_{na}/N_{nb}) = \log(N_{oa}/N_{ob}) + n \cdot \log(\mathit{eff}_a/\mathit{eff}_b) \qquad [2]$$

In this case, the initial concentrations of the two isoforms have to be deduced separately from the yields measured after various numbers of cycles, extrapolating back to 0 cycles (17, 18), or from co-amplification of standards that used the same primers with the same efficiency. Fortunately, it seems that the efficiencies for amplification from isoforms with common primers are often similar. However, pilot experiments should be performed in which a $5'$-^{32}P end-labelled primer (see *Protocol 2*) is used for the RT-PCR reactions and the yield of isoforms is measured after different numbers of cycles. Plot log N_n against *n* and determine *eff* from the slopes. If material is scarce, then a range of known concentrations of cDNA or, better still, RNA derived from cloned isoforms may be used. If the efficiencies are the same, then it is reasonable to measure the ratio of two products from one reaction with the test sample even without checking that the reaction is still exponential, as long as the total yields are less than those at which the reaction efficiency was found to decline in the pilot experiments.

The decline in reaction efficiency with excessive numbers of cycles particularly affects the relative proportion of longer products. This can be attributed to depletion of dNTPs once a certain level of product has formed; depletion seems to cause a failure to complete the extension of longer products. Obviously, this is less significant with isoforms of equal distance between the primers, and pilot experiments in which the starting substrates are mixed at various known ratios may show that the ratios found after overamplification are still valid. In addition, reverse transcription itself is prone to underrepresent the proportion of longer forms.

Protocol 4 describes the basic strategy for RT-PCR (19). The PCR primers should be chosen so that the shortest and longest products can be easily fractionated on a suitable gel or sequenced. The primers should contain a $5'$ extension with a convenient restriction enzyme cleavage site (for subsequent cloning); $5'$ to this include 8-10 nt of arbitrary sequence, GC-rich, but unable to form strong secondary structures. The $3'$ end of the primer should not end in G or C. The ~20 nt at the $3'$ ends of the primers should match or be complementary to mRNA exon sequences. If there are corresponding endogenous transcripts in the cells used for transfection, then at least one of the two primers should be directed to exon sequences which derive from the vector or from non-specific sequences inserted in the exogenous gene. It is helpful if the two primers have approximately the same calculated T_m.

3: Splicing mRNA precursors

Protocol 4. Reverse transcription and amplification by PCR[a]

Equipment and reagents

- 10×PCR buffer, *Taq* polymerase, and PCR apparatus (see *Protocol 2*)
- Oligonucleotide primers (>10 pmol/µl) chosen as described in the text; the RNA-sense primer could have been labelled, as in *Protocol 2*, to facilitate subsequent analysis
- Test RNA, total RNA (5 µg per reaction) or poly(A)+RNA, see Chapter 1 (*Protocols 9–12*) for preparation
- RNase inhibitor, e.g. Inhibit-Ace (5 prime-3 prime Inc.), RNAguard (Pharmacia), or RNasin (Promega)
- Reverse transcriptase from Moloney murine leukaemia virus (16 units/µl; Pharmacia)[b]

Method

1. Prepare each 10 µl reaction in a 0.5 ml microcentrifuge tube. Add 10 pmol of the specific downstream oligonucleotide primer, 5 µg of total RNA, or the equivalent amount of poly(A)+RNA, 1 µl of 10×PCR buffer, and water to 10 µl. Include one reaction with the RNA replaced by water.
2. Heat the samples at 90 °C for 5 min to denature the RNA, then anneal them at 65 °C for 5 min.
3. Add to each tube 20 units of RNase inhibitor (0.5 µl) and 10 units of Moloney murine leukaemia virus reverse transcriptase. Incubate the samples at 37 °C for 1 h.
4. Prepare *Taq* mixture. For each reaction, mix 1 µl of 10×PCR buffer, 10 pmol upstream oligonucleotide primer, 1 unit of *Taq* polymerase, and H_2O to 10 µl.
5. Add 10 µl of *Taq* mixture to each reaction (from step 3). Overlay each reaction with paraffin oil and amplify. Conditions for the PCR amplification are given in *Protocol 2*. However, the optimal conditions may have to be determined in a pilot experiment as described for *Protocol 2*. An easier alternative is to use a 'touchdown' programme (20), in which the annealing temperature is reduced progressively by 2 °C, with two cycles at each temperature except the lowest, at which there are 25 cycles. This procedure usually produces clean results for any PCR amplification, with a much reduced incidence of non-specific products. It does not require the determination of optimum conditions for each combination of primers and template, and substantially reduces the importance of ensuring both that the two primers will have comparable values of T_m and that there are no other likely sites of annealing to the template.

[a] Based on ref. 19.
[b] A procedure using a thermostable enzyme is described in Chapter 2 (*Protocol 6*). *Taq* DNA polymerase has some reverse transcription activity.

After PCR amplification (*Protocol 4*), several methods of analysis are available. The quality of the PCR can be checked by direct electrophoresis of a portion of the reaction on an agarose gel (see Chapter 1, *Protocol 6*). For direct analysis of the ratios of products of various lengths, the RNA-sense primer could be 5′ end-labelled with [γ-^{32}P]ATP and polynucleotide kinase and a portion of the PCR reaction products fractionated directly on a gel, as described in *Protocol 2*. The yields can be measured either by direct radioisotope imaging, laser densitometry of autoradiographs, or excision of the bands and scintillation counting. If the isoforms produce PCR fragments of the same length, it may be possible to alter the length of one PCR product selectively by cleavage of the RT–PCR product mixture with a restriction enzyme prior to gel electrophoresis. If the autoradiographic signals are too faint, or there is a high background of non-specific amplification, then a small portion (10^{-8}–10^{-2}) of the first amplification can be added to a new reaction mixture in which one (or both) of the primers is different and so would prime at a new site within the first PCR product. Again, one of the primers in such a 'nested' PCR can be labelled to facilitate analysis. It might be appropriate to minimize the number of cycles to keep the PCR reaction within the exponential phase.

Statistically significant results can also be obtained by cloning and sequencing sufficient PCR products (21), which has two advantages over direct fractionation: the exons in any products of unexpected length can be identified and products of identical length, but different in sequence, can be distinguished. However, it has the disadvantage that the different isoforms may not be cloned with equal efficiencies. This approach is only realistic if a method is used for preparing large numbers of sequencing templates simultaneously. The PCR products should be purified and cloned into M13. A method has been developed for preparing M13-type templates in batches of 96, allowing very large numbers to be prepared in a day (22). The procedure involves two departures from normal preparations: filtration is used to separate bacteriophage from bacteria, and phenol extraction is not used (22, 23). This procedure and variations upon it have been reviewed elsewhere (24). The method given in *Protocol 5* is the one used currently in the laboratory of one of the authors (Ian Eperon) and incorporates the culture conditions described in ref. 24. Growth of the phage, preparation of the DNA templates, the sequencing reactions, and the gel electrophoresis can all be done within one day. To assess the relative proportions of different classes of splicing event, the DNA templates prepared in *Protocol 5* are sequenced with only one ddNTP in numbers sufficient to give statistically useful results.

Protocol 5. Preparation of 96 templates for DNA sequence analysis

Equipment and reagents

- A centrifuge with a rotor suitable for microtitre plates. Microtitre plates can be centrifuged at 600 g without any problems. If centrifugation at 2800 g is desired (see steps 7 and 12), support the base of the wells, otherwise the plate will crack because it sits on the thin bottom rim which protrudes below the base of the wells. The support can be a plate from which the rim has been removed, or a flat solid plate of any material which fits within the rim and protrudes below it

Continued

3: Splicing mRNA precursors

Protocol 5. *Continued*

- Repetitive pipettors (essential), and a 96-tip automatic pipette (useful but not essential, e.g. Handi-Spence II, Sandy Spring Instrument Co. Inc., or Costar Transtar-96)
- 8-well filter strips (0.2 µm cellulose acetate; Costar, cat. no. 8511) with 96-well frame
- A 96-well filter strip spacer (Costar)
- Adhesive sealing film for microtitre plates (Titertek non-toxic plate sealers; Flow Labs)
- A dense (overnight) culture of *E. coli* JM101
- PEG–NaCl (20% polyethylene glycol 6000, 2.5 M NaCl)
- TES buffer (10 mM Tris–HCl, pH 7.5, 0.1 mM EDTA, 0.5% SDS)
- TE buffer (10 mM Tris–HCl, pH 7.5, 0.1 mM EDTA)
- Microtitre plates (plastic, flat-bottomed, 96-well; Nunc)
- M13 clones from RT–PCR procedure (see *Protocol 4*), as plaques on agar plates
- Wooden toothpicks
- 37 °C shaking incubator
- 95% ethanol, 0.12 M sodium acetate, pH 5–6
- Bacterial growth medium (e.g. 2 × TY broth)

Method

1. Dilute the culture of *E. coli* JM101 20-fold with fresh growth medium and add 0.3 ml aliquots into as many wells of a 96-well flat-bottomed microtitre plate as required.
2. Inoculate each well with a separate M13 plaque, using a fresh toothpick for each.
3. Shake the plate for up to 5 h at 37 °C at 300 r.p.m. Wedging the open plate (open to prevent oxygen depletion) in a small closed box will prevent any losses by evaporation. The box is, in turn, fixed firmly on to the platform of the shaker (e.g. by wedging it in between flask holders).
4. Insert 8-well filter strips into the frame. Place this on top of the 96-well filter strip spacer, above a new 96-well microtitre plate. Transfer the cultures from the growth plate into the filter wells with an automatic pipette (preferably multitip).
5. Filter the cultures by centrifugation at 300–600 g. The cells will be retained, and the supernatant will pass into the new plate.
6. Add ~40 µl of PEG–NaCl to each well. Cover the plate with adhesive film and invert it to mix the contents of the wells.
7. Centrifuge the plate at 2800 g for 5 min, or 600 g for 20 min, to recover the bacteriophage as pellets. Remove the supernatants by aspiration using a vacuum line with trap.
8. Briefly centrifuge the plate to deposit the residual liquid at the bottom of the wells. Remove by aspiration.
9. Add 40 µl of TES buffer to each well. Resuspend the phage by vortexing.
10. Incubate the plate by floating it on water at 80 °C for 5–10 min.
11. Add 100 µl of 95% ethanol, 0.12 M sodium acetate to each well. Do not chill.

Continued

Protocol 5. *Continued*

12. Centrifuge the plate at 2800 g for 10 min (or 600 g for 20 min). Remove and discard the supernatants by aspiration.
13. Wash the pellets with 95% ethanol. Centrifuge briefly, remove the supernatants, and allow the plate to air dry.
14. Dissolve each DNA pellet in 20–50 μl of TE buffer.
15. Transfer 2 μl of each DNA template to a fresh microtitre plate for screening by sequencing.

3. Analysis of RNA splicing in metazoan nuclear extracts

3.1 Preparation of active extracts

3.1.1 Choice of cells

Splicing of exogenous transcripts was first reported after microinjection of RNA into the nuclei of *Xenopus* oocytes (25). Subsequently, several procedures were developed for preparing active splicing extracts from HeLa cells grown in suspension (26–28). Of these, the one based on the method of Dignam *et al.* (29) for making active transcription extracts from nuclei was the most efficient (26), and it has been the cornerstone of most subsequent work on splicing mechanisms. The method involves hypotonic swelling of the cells, disruption of the plasma membrane, collection of the nuclei, and precipitation of the chromatin with high salt. Similar methods have been used to prepare extracts from other cells in suspension, such as Namalwa cells (30) and 293 cells (31), and from F9 mouse teratocarcinoma cells, which adhere poorly to surfaces and can be dislodged easily (32). A small-scale version has been described for the preparation of extracts from as few as 3×10^7 mammalian cells grown as a monolayer (33). This was reported first with HeLa cells (33), and it has been used for a variety of fibroblast-based cell lines (34). The same principles have been used for the preparation of extracts from both Kc cells and 0–12 h *Drosophila* embryos (35) and from fertilized *Ascaris* eggs at the 32-cell stage (36).

Few of the extracts prepared from alternative cell lines are as active or as problem-free in splicing assays as those from HeLa cells. Recent results with fibroblast nuclear extracts have suggested that proteolysis of splicing factors was a problem, which was partly avoided by using protease inhibitors or cells transformed with SV40 large T antigen (34). Studies of regulated, alternative splicing are likely to require extracts from unfavourable cell lines. In this case, it might be possible to use these extracts as supplements to the basic HeLa cell

3: Splicing mRNA precursors

extracts. If the tissue- or cell-specific pattern of splicing is produced, this can form the basis of any assay for purification of the activity. Several examples of this approach have been described (32, 37, 38).

3.1.2 Preparation of HeLa cell nuclear extracts

HeLa cells are grown in suspension in DMEM or similar medium. Although frozen cell stocks must be resuscitated in medium containing 10% fetal calf serum, this can be replaced by 10% newborn calf serum when the cells are established. It is absolutely essential that the cells should be thriving, with a very low proportion of dead cells. The cell population should double every 24 h. Standard spinner cultures should be harvested at a density of $5 \times 10^5 - 1 \times 10^6$ cells per ml. Preparation of a HeLa cell splicing extract by the Dignam method is described in *Protocol 6*; this requires 2×10^9 cells.

Protocol 6. Preparation of a HeLa cell nuclear extract[a]

Equipment and reagents

- Dounce homogenizer (glass, 15 ml size with tight-fitting pestle)
- PBS (20 mM potassium phosphate, pH 7.4, 0.13 M NaCl)
- PMSF stock solution (0.13 g PMSF in 7.5 ml propan-2-ol). Make just before use
- 0.5 M EDTA (potassium salt, pH 8.5 with KOH)
- Buffer A base (10 mM KCl, 1.5 mM MgCl$_2$)
- Buffer C base (20 ml glycerol, 4.5 g KCl, 40 µl 0.5 M EDTA, and H$_2$O to 97.5 ml)
- Buffer D base (44.7 g KCl, 0.3–1.2 litres of glyccrol[b], 2.4 ml 0.5 M EDTA, and H$_2$O to 5.85 litres); autoclave this solution at 120 °C for 15 min (pressure of approx 1.08 kg/cm^2) in aliquots of 975 ml
- 1.0 M Hepes–KOH, pH 7.9[c]. Sterilize this solution by filtration through a 0.2 µm membrane filter (see *Protocol 1*)
- 1.0 M DTT. Store at −20 °C
- Buffer A (50 ml of buffer A base, 0.5 ml of 1.0 M Hepes–KOH, 25 µl of 1 M DTT). Prepare buffer A fresh, and keep it on ice
- Buffer C (10 ml of buffer C base, 0.2 ml of 1.0 M Hepes–KOH, 50 µl of PMSF stock solution, 5 µl of 1.0 M DTT). Prepare buffer C fresh, and keep it on ice
- Buffer D (975 ml of buffer D base, 20 ml of 1.0 M Hepes–KOH, 5 ml of PMSF stock solution, 0.5 ml of 1.0 M DTT). Prepare buffer D fresh, and keep it on ice
- Dialysis clips and tubing (the tubing is autoclaved in water and stored at 4 °C)
- 50 ml polypropylene centrifuge tubes (Oak Ridge style), 1 litre centrifuge bottles and 50 ml screw-capped disposable tubes
- HeLa cell spinner culture (2×10^9 cells in 4 litres of DMEM plus 10% newborn calf serum)

Method

NB. Unless otherwise stated, use precooled centrifuges, rotors, and solutions.

1. Sterilize the dialysis clips, polypropylene centrifuge tubes, and the homogenizer by autoclaving at 120 °C for 15 min (approx. 1.08 kg/cm^2).

Continued

Protocol 6. *Continued*

2. Decant the cells into 4 × 1 litre bottles. Centrifuge at 75–300 g (maximum), for 10 min.
3. Pour away the supernatant and resuspend each cell pellet in 20 ml of PBS. Combine the suspension in two 50 ml disposable tubes on ice.
4. Centrifuge the cell suspension at ~450 g (max.) for 5 min. Resuspend and pool the cells in 30 ml of buffer A.
5. Centrifuge the cells, as in step 4, and resuspend them in 14 ml of buffer A.
6. Decant the suspension into the homogenizer on ice. Disrupt the swollen cells with about 10 strokes of the tight pestle. Decant the homogenate into a 50 ml polypropylene centrifuge tube. Rinse the homogenizer and pestle in buffer C.
7. Collect the nuclei by centrifugation in a precooled rotor at 4300 g for 15 min.
8. Remove the supernatant by cautious aspiration and then centrifuge the tube again at 26 000 g for ~15 min. Remove and discard the supernatant.
9. Use buffer C to transfer the nuclear pellets to the homogenizer. A total of 4 ml buffer C is required.
10. Homogenize the nuclei with 10 strokes of the tight pestle.
11. Transfer the nuclear homogenate to a polypropylene centrifuge tube and allow it to rotate gently for 30 min at 4 °C.
12. Centrifuge the extract at 26 000 g for ~30 min.
13. Transfer the supernatant to dialysis tubing and dialyse it at 4 °C for several hours against changes of buffer D (in total, 1 litre of buffer D).
14. Centrifuge the extract again at 26 000 g.
15. Decant the supernatant into a tube, mix it gently and divide it into small aliquots. Store them at −70 °C. The extract is stable indefinitely.

[a] Based on refs 26 and 29.
[b] The concentration of glycerol in buffer D can vary from 5–20%, with no significant effect on the activity or stability of the nuclear extract. Solutions containing the lower concentrations of glycerol are easier to pipette. It has been reported that, at least with some RNA substrates, splicing is more efficient if buffer D contains 80 mM monopotassium glutamate instead of 0.1 M KCl (39).
[c] 1 M Hepes is normally used, but it has been reported that extracts made using 1 M triethanolamine–HCl, pH 7.9, are more efficient (40).

3.2 Splicing reactions *in vitro*

3.2.1 Choice of RNA substrate

There are several points which may help in designing the substrate:

- If the template is prepared by PCR, then the 5′ and 3′ termini of the transcript can be altered at will.

3: Splicing mRNA precursors

- Incorporation of a 5'-terminal 'cap' blocks 5' exonuclease activity in the extract (26, 41).
- The 3' terminus can be stabilized against RNase activity by incorporating a 3' hairpin of 10–12 bp when the anti-sense PCR oligonucleotide is designed (see Chapter 1, Section 2.1).
- Incorporating a 5' splice site at the 3' terminus of the 3' exon increases splicing efficiency (42).
- Shorter introns are spliced more efficiently.

3.2.2 Standard splicing reactions

The procedure described in *Protocol 7* is for splicing reactions in polypropylene tubes 0.5 ml or 1.5 ml in size. The products of splicing *in vitro* are separated by electrophoresis on a sequencing type gel; in the laboratory of one of the authors (Ian Eperon), 20% formamide is included in the gel as well as 7 M urea to reduce the chance that secondary structure will cause aberrant mobility of the RNA. The percentage of acrylamide can be varied to optimize ease of detection of the splicing intermediate and products. Lariat molecules with loops of up to ~ 150 nt migrate like linear molecules on gels of 4% or 5% acrylamide, and more slowly on more concentrated gels. Lariat molecules with loops of about 600 nt or more are very markedly retarded even at low acrylamide concentrations.

Protocol 7. Standard assays of splicing *in vitro*[a]

Equipment and reagents

- Nuclear extract and buffer D prepared as in *Protocol 6*
- ^{32}P-RNA substrates (~ 1 fmol/μl) prepared by transcription *in vitro* from a suitable template (see Chapter 1, *Protocol 4*)
- 80 mM MgCl$_2$
- 0.1 M ATP
- 0.5 M phosphocreatine, sterilized by filtration through a 0.2 μm membrane filter
- 13% polyvinyl alcohol (PVA)
- RNase inhibitor (see *Protocol 4*); the volume given is for RNAguard (Pharmacia)
- Proteinase K (10 mg/ml in water); store this in aliquots at – 20 °C
- Proteinase K buffer (2 ml of 1.0 M Tris–HCl, pH 7.5, 0.5 ml of 0.5 M EDTA, 0.6 ml of 5.0 M NaCl, 2 ml of 10% SDS, 14.9 ml water)
- Phenol
- Materials for denaturing polyacrylamide gel electrophoresis (see Chapter 1, *Protocol 8*)

Method

1. In a microcentrifuge tube on ice, make up a splicing reaction mixture as follows:
   - nuclear extract   125 μl
   - RNase inhibitor   9 μl (300 units)

Continued

Protocol 7. *Continued*

- 80 mM MgCl$_2$ 12 µl
- 0.5 M phosphocreatine   12 µl
- 0.1 M ATP   5 µl
- 13% PVA   60 µl
- buffer D   to 295 µl final vol.

The final volume of 295 µl will suffice for about 24 additions of 10 µl, the losses being caused by the viscosity of the mixture.

2. For 4 time points in the splicing reactions, mix 2.5 µl (~2.5 fmol) RNA with 50 µl of reaction mixture (step 1) on ice[b].

3. At predetermined times over 2–4 h, place an aliquot of 10.5 µl into a microcentrifuge tube and incubate at 30 °C, such that all the time course samples finish their incubations together.

4. Prepare the proteinase K mixture by adding 1 vol. of proteinase K stock solution to 100 vol. of proteinase K buffer. Each time course sample (step 3) requires 200 µl of this mixture.

5. Add 200 µl of the proteinase K mixture to each completed splicing reaction and incubate all the tubes at 37 °C for 15 min.

6. Remove the proteins by vortexing each sample with an equal volume of phenol. Centrifuge for 5 min in a microcentrifuge to separate the phases. Avoiding the interface, remove the upper (aqueous) layer from each sample into a fresh tube.

7. Add 0.5 ml of absolute ethanol to each tube to precipitate the RNA. Do not cool the samples.

8. Recover the RNA precipitates by centrifugation in a microcentrifuge for 15 min. Rinse the pellets with 95% ethanol and allow them to air dry.

9. Analyse the RNA products by electrophoresis on a denaturing polyacrylamide gel followed by autoradiography (see Chapter 1, *Protocol 8*).

[a] Based on ref. 26.
[b] Note that if the time course of *association* of splicing factors with the substrate RNA is to be analysed rather than *splicing*, then the substrate should be mixed with the extract only at the start of each incubation. Nevertheless, it is still often helpful to ensure that all incubations end at the same time (step 3).

The band patterns produced on electrophoresis of splicing reaction products can be hard to interpret, particularly if there are several possible alternative splicing reactions or multiple introns in the substrate. The main guides to interpretation are the time course of the appearance of each band and its electrophoretic mobility. Intermediates appear first and may decline when the final products form, and those molecules which include the 3′ end of the transcript are usually slightly more heterogeneous in size than the 5′ exon intermediate and lariat product. More rigorous assignments can be made by using RNase H or reverse transcriptase to map the RNA bands. In addition, lariats can be detected by looking for altered relative mobilities

3: Splicing mRNA precursors

after electrophoresis on gels of different polyacrylamide concentrations. A more complete description of these methods is given in Chapter 2 (Section 4).

The reaction conditions used for splicing were optimized for the first intron of the human β-globin gene (26). However, the concentration of the nuclear extract and the ionic strength of the final splicing reaction can influence very substantially the choice among alternative splice sites (43–45). In some cases it has been possible to study natural examples of alternative splicing only after adjusting these conditions (37, 39, 46, 47).

3.2.3 Splicing reactions in microtitre plates

It is often helpful to be able to deal with a number of substrates or to extract preparations simultaneously. For this reason, a method has been developed for handling splicing reactions in microtitre plates (refs 45, 103). This makes the task so easy that microtitre plates have wholly replaced single tube reactions for studying splicing in I. Eperon's laboratory. In essence, *Protocol 7* is followed except that the phenol:chloroform extraction is omitted.

The procedure using microtitre plates is described in *Protocol 8*. Two forms of microtitre plate work well: standard-size polystyrene plates with either flat or V-shaped bottoms to the wells. Plates from some commercial sources may cause a slight increase in RNA degradation, but this is readily eliminated by siliconizing the plates. The siliconized surface also has advantages when multiple transfers are required, as in RNase H protection experiments (Section 3.2.4).

Protocol 8. Splicing reactions in microtitre plates

Equipment and reagents
- Microtitre plates (96-well, polystyrene plates with flat-bottom or V-shaped wells able to hold ~ 300 μl); if necessary, they can be siliconized by dipping them quickly into a solution of dimethyldichlorosilane and dried inverted in air
- Nuclear extract and buffer D prepared as in *Protocol 6*
- RNA substrate, 80 mM $MgCl_2$, 0.1 M ATP, 0.5 M phosphocreatine, 13% PVA,
- RNAse inhibitor, proteinase K and proteinase K buffer (see *Protocol 7*)
- Microtitre plate centrifuge, pipettors (repetitive and 8-channel; not essential) and plate-sealing film (see *Protocol 5*)
- Polyacrylamide gel electrophoresis equipment, and formamide–dye mixture (Chapter 1, *Protocol 8*)

Method

1. Assemble the splicing reaction mixture as in step 1 of *Protocol 7*. Add the RNA substrate to generate the RNA-reaction mixture (*Protocol 7*, step 2). The volume required will depend on the number of time course samples required (steps 2, 3, and 5). If there are multiple substrates

Continued

Protocol 8. *Continued*

  or extract preparations to be tested, then these mixtures can be placed in a microtitre plate on ice, such that the subsequent withdrawal of aliquots to initiate incubation for each time point (step 2) can be done with a multichannel pipettor.

2. Transfer 10.5 µl portions of the reaction mixtures requiring the longest incubation time to the wells of a microtitre plate. Seal the plate with adhesive film and place it in an incubator at 30 °C. If the incubator is warmer near the bottom, invert the plates to eliminate volume losses by evaporation and condensation on to the seal. Surface tension holds the droplet in place in the well.
3. Add further samples to new wells in the plate at appropriate times.
4. Prepare the proteinase K mixture by adding 1 vol. of proteinase K stock solution to 25 vol. of proteinase K buffer. Each sample requires 50 µl.
5. Complete the incubation then add 10.5 µl portions of the unincubated reactions to empty wells and add 50 µl of the proteinase K mixture to each sample. Seal the plate and incubate at 37 °C for 5 min.
6. Add 150 µl of 95% ethanol to each well. Immediately centrifuge the plate, without cooling, at 2800 g for 10 min (or 600 g for 20 min).
7. Remove the supernatant by aspiration. Add 200 µl of absolute ethanol per well (this volume saves re-balancing the plate for centrifugation). Centrifuge for 5 min as in step 6.
8. Aspirate the supernatants and allow the pellets to air dry.
9. Dissolve each sample in 10 µl of the formamide–dye mixture, vortexing vigorously to wet the entire base of the wells. Analyse the RNA by denaturing polyacrylamide gel electrophoresis (see Chapter 1, *Protocol 8*).

3.2.4 Analysis of splicing using RNase H

The association of some factors with RNA during splicing may be analysed using procedures based on the enzymatic activity of RNase H. RNase H digests RNA in RNA:DNA hybrids but not RNA alone, which allows the investigator to use synthetic deoxyoligonucleotides to direct RNase H cleavage of specific RNA sequences (Chapter 2, Section 3.3). Some RNA sequences containing splicing signals become resistant to challenge by oligonucleotide-directed RNase H cleavage as splicing components assemble (48, 49, 103).

The use of siliconized V-well microtitre plates lends itself very well to studies of the protection of splicing signals against oligonucleotide-directed RNase H cleavage. It is easy to process many samples simultaneously and at defined, short invervals of time, maximizing the reliability and reproducibility of comparisons between different substrates or extracts (103). The steps involved are shown diagrammatically in *Figure 3* and are described in *Protocol 9*.

3: Splicing mRNA precursors

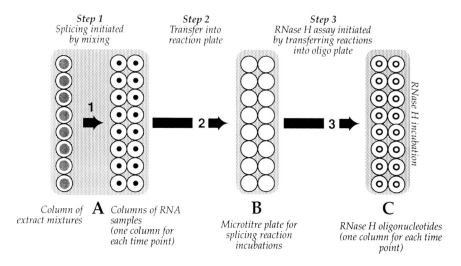

Figure 3. Use of microtitre plates in RNase H protection assays. A typical 96-well (8 × 12) microtitre plate can be regarded as having 8 horizontal *rows* of wells (each comprising 12 wells) and 12 vertical *columns* of wells (each comprising 8 wells). This allows up to 8 splicing reaction time courses to be followed simultaneously and rapidly. The splicing reaction mixture (with added RNase H, but no RNA) is dispensed into the wells of one column in plate A. RNA samples are placed into the wells of adjacent columns, and replicate aliquots are added to successive columns until the number of wells containing RNA in each row matches the number of time points to be analysed. Plate A is kept on ice. For each time point, 10 µl of extract is removed simultaneously from each of the wells in the extract column using an 8-channel pipettor, expelled into the wells of one column of RNA samples and immediately taken back into the pipettor (step 1). The mixed RNA–extract is then expelled (step 2) into the wells of a column in a fresh microtitre plate (B) for incubation, thus initiating the splicing reactions. The process is repeated for the next time point, returning to the same extract wells, but mixing with the next column of RNA samples and transferring them to a new column in plate B. Samples are timed so that all the reactions in plate B finish together. Target oligonucleotides for the RNase H reaction are likewise dispensed in columns in plate C. When all the reactions (B) are ready, each column in plate B is mixed with oligonucleotide in a corresponding column of plate C (step 3) and the RNase H reaction allowed to take place. For controls (no incubation at 30 °C), the substrate RNA encounters the extract and the oligonucleotide simultaneously. Aliquots of extract from plate A are mixed directly with oligonucleotide (plate C), transferred back to a column of RNA in plate A and finally returned to plate C for RNase H digestion.

Protocol 9. RNase H protection assays with splicing in microtitre plates

Equipment and reagents

- 8-channel pipettor able to pipette volumes of 5–10 µl (e.g. Flow Titertek, 5–50 µl digital)
- Microtitre plates with V-shaped wells (see *Protocol 8*)

Continued

Protocol 9. *Continued*

- RNA substrate(s), 80 mM MgCl$_2$, 0.1 M ATP, 0.5 M phosphocreatine, 13% PVA, RNase inhibitor, proteinase K, and proteinase K buffer (see *Protocol 7*)
- RNase H (~0.8 units/µl; Pharmacia)
- Oligonucleotides (14mers) complementary to the target sequences (50 pmol per µl in buffer D; *Protocol 6*)

Method

1. Assemble the splicing reaction mixtures as in step 1 of *Protocol 7*. Do not add the RNA substrate, but add 8 µl (6.4 units) of RNase H per 295 µl of the mixture. Divide the mixtures into the wells of a single column of a microtitre plate, with one well for each RNA substrate or reaction condition to be tested (Plate A in *Figure 3*). Calculate the number of time-course samples to be performed with each portion of the reaction mixture, and ensure that the volume of extract dispensed into each well is at least 125% of the calculated volume required. Keep plate A on ice.
2. For each portion of reaction mixture in a well in the first column of the microtitre plate, dispense 0.5 µl aliquots of RNA into further wells along the same row of the microtitre plate, the number of aliquots in each row being the number of time-course samples to be analysed (see *Figure 3*).
3. Use an 8-channel pipettor to transfer and mix 5–10 µl of each reaction mixture in the first column into one of the columns of RNA samples (*Figure 3*, step 1). Transfer the mixtures to a column of a new plate (*Figure 3*, plate B, step 2).
4. Incubate plate B at 30 °C as described in *Protocol 8*. These are the samples receiving the longest incubations.
5. At appropriate intervals, repeat step 3 with other sets of RNA aliquots, transferring the reactions to the next columns in plate B.
6. Dispense 2 µl aliquots of the required oligonucleotide solutions into the wells of a third plate (plate C) in an array matching that in plate B.
7. About 2 min before the end of the incubation time, start control reactions (0 min at 30 °C) by transferring and mixing the standard volume of each reaction mixture with a column of oligonucleotides in plate C.
8. Transfer these oligonucleotide–extract mixtures to the last column of RNA samples in plate A.
9. Return the oligonucleotide–RNA–extract mixtures to the original wells of plate C.
10. At the end of the incubation period, mix each column of timed reactions in plate B (beginning with the shortest time, for maximum accuracy) into a column of oligonucleotides in plate C (*Figure 3*, step 3).
11. Incubate the oligonucleotide plate at 30 °C for 15 min.
12. Treat the samples with the proteinase K mixture and analyse the samples by electrophoresis as described in *Protocol 8*, steps 4–9.

3.3 Depletion of components from nuclear extracts

The first method used for removal of specific components was immunodepletion, which used antibodies directed against snRNPs (50). However, it was not possible then to restore splicing activity by supplementation with the deleted component. Several recent reports have described the depletion of specific protein components followed by restoration of activity. In one case, nuclear extract in a high-salt buffer (0.42 M NaCl, ref. 29) was incubated with antibodies bound to Protein A–Sepharose, and the antigen removed by centrifugation (51). The high-salt concentration prevented the removal of other proteins associated with the antigen. In another procedure, nuclear extract was pre-incubated with ATP to release the antigen from aggregates. The antigen was then loaded on to an antibody-affinity column from which excess buffer had been removed by centrifugation. After removal of the antigen, the depleted extract was recovered by centrifugation (52, 53).

A common strategy for selective inactivation of snRNPs in extracts is the use of oligonucleotide-directed RNase H digestion of exposed portions of snRNA (50, 54, 55). Splicing mixtures are prepared as described in step 1 of *Protocol 7*, with RNase H added as in *Protocol 9*, step 1. The mixtures are incubated at 30 °C for 45 min with ~700 pmol of the specific oligonucleotide per 300 µl mixture. This oligonucleotide–extract mixture is then used for splicing as in *Protocols 7–9*. The amount of oligonucleotide required depends on the efficiency of the cleavage and the extent of any cross-reaction when the substrate is added. An indication of the extent of cleavage and its efficiency can be gained by electrophoresis of a portion (10 µl) of the reaction on a denaturing polyacrylamide gel, followed by ethidium bromide staining. A more rigorous analysis requires Northern blotting or primer extension on RNA isolated from samples both before and after splicing (see Chapters 2, 4, 6 for descriptions of these techniques).

Oligonucleotide-directed inactivation of snRNPs has been used very effectively with *Xenopus* oocytes for studies on the role of snRNPs and the interactions among their RNA components. In principle, micro-injected oligonucleotides should cause degradation of endogenous snRNPs, allowing an exogenous (mutated) snRNA analogue to be injected and its activity to be tested with a micro-injected pre-mRNA substrate. In practice, the best route was found to involve injection of the oligonucleotide along with a gene encoding the mutant U snRNA, the splicing substrate being introduced later by injection of labelled RNA substrate or the corresponding gene (56, 57).

The most rigorous procedure described for the depletion of snRNPs involves the use of biotinylated oligonucleotides which hybridize to exposed portions of the snRNAs and are removed by streptavidin–agarose. These oligonucleotides comprise 2'-*O*-alkyl ribose derivatives, which confer stability and do not attract RNase H activity (58, 59). These methods are described in more detail in Chapter 4 (Sections 3 and 4).

Ian C. Eperon and Adrian R. Krainer

4. Fractionation of splicing extracts and biochemical identification of splicing factors

This section describes general strategies for the fractionation of extracts, discusses practical points that should be taken into account when designing a purification scheme, and reviews methods that have been used for the identification and purification of specific splicing factors.

4.1 Biochemical fractionation strategies

4.1.1 General considerations

Although a great deal of information about splicing mechanisms has been obtained by studying the reaction in crude extracts, it is necessary to identify, purify, and characterize individual factors to gain a detailed understanding of the molecular mechanisms of this complex reaction. The fractionation of splicing extracts into individual components is a complex task, particularly since the total number of constituents is not known, and may be very large. At present, it is known that the snRNPs U1, U2, U4/U6, and U5 are essential factors for splicing in both metazoan and yeast cells. Several additional protein factors are also necessary, some of which have been identified biochemically in mammalian systems and genetically in yeast systems (reviewed in refs 60–62).

In the biochemical fractionation of a complex system, the immediate objective is to separate the crude extract into two or more fractions, each of which is required for complete splicing. When multiple components are involved, this process can be sequential, so that each fraction is then subfractionated further. However, in practice, some of the factors may be present in several fractions, which greatly complicates the analysis. A strategy, that partially circumvents this problem, is to devise a way to remove or inactivate specifically a single component from the crude extract. The resulting depleted extract is a convenient source of all splicing factors except the one to be purified. A second untreated extract is used to purify the component of interest, and complete splicing is obtained by mixing the depleted extract and the partially or completely purified component. Thus, the depleted extract is used for the purification assay as well as for the subsequent functional characterization of the purified component. This approach has been successful for human splicing factors SF2 (ref. 63), SC35 (ref. 52), U2AF (ref. 64), and the 88 kDa polypeptide recognized by monoclonal antibody 53/4 (ref. 51).

A common problem in purifying active splicing factors is that many of these appear to be present in limiting concentrations. Thus, the accumulated losses inevitably associated with the multiple steps of a conventional separation scheme may make purification almost impossible. Furthermore, crude splicing extracts are rarely sufficiently active to withstand the degree of dilution commonly associated with conventional procedures. These problems are compounded if multiple components are fractionated simultaneously.

Splicing reactions have very specific buffer requirements, in particular the optimal concentration ranges for monovalent and divalent cations are narrow and depend on the pre-mRNA substrate used. After most purification steps, it is necessary to alter the ionic composition of the fractions prior to assay. It is seldom possible simply to dilute a fraction to lower its salt concentration, unless it is highly active to begin with. Dialysis is usually employed. When dealing with multiple fractions, it is convenient to dialyse small aliquots of each fraction in a multiple dialysis apparatus (BRL, Pierce, Spectrum). If fractions need to be concentrated, this can be done by ultrafiltration in centrifugal units (Amicon) or by precipitating the protein with 80–90% ammonium sulphate.

4.1.2 Preparation of starting extracts

Extracts prepared from cultured cells by the method of Dignam *et al.* (29) are the usual starting material. Section 3 summarizes some of the background to this method and *Protocol 6* describes the preparation of standard-sized extracts suitable for analysis of splicing. However, biochemical purification of factors present in limiting quantities usually requires large amounts of starting material. Tissue culture cells that grow in suspension are readily scaled up although this is expensive and considerable time is required to maintain larger spinner cultures as well as to harvest the cells. Fortunately, active splicing extracts can be prepared from frozen cells, so smaller, more manageable cultures may be grown up over a period of time and the cells stock-piled. *Protocol 10* describes the preparation of extracts from large amounts of frozen HeLa cells ($2-5 \times 10^{10}$ cells or more). Successful purification of active splicing factors from animal tissues has not yet been reported.

Protocol 10. Preparation of HeLa cell extracts on a large scale[a]

Equipment and reagents

- HeLa cell spinner cultures. Grow the cells to 10^9 cells/litre in DMEM + 5% calf serum; a total of at least $2-5 \times 10^{10}$ cells is required
- Buffer A and PBS (see *Protocol 6*)
- 50 ml disposable polypropylene tubes with screw caps and 1 litre centrifuge bottles
- All other materials and reagents described in *Protocol 6*

Method

1. Harvest the cells by centifugation (see *Protocol 6*, step 2).
2. Discard the medium and resuspend the cell pellets in 50 ml of cold DMEM containing 50% glycerol per 3 litres of original culture. Store the suspensions in 50 ml polypropylene tubes at −70 °C until required.
3. Place the frozen pellets at 30 °C until just thawed.
4. Centrifuge the cell suspension at 1800 g_{max} for 20 min at 4 °C.

Continued

Protocol 10. Continued

5. Discard the supernatant and resuspend each pellet in 30 ml of cold PBS.
6. Pool the cell suspensions and centrifuge as in step 4. The cells will swell upon removal of the glycerol.
7. Discard the supernatant and resuspend the pellet in an equal volume of buffer A.
8. Homogenize the cells, prepare, and store the extract as described in *Protocol 6*, steps 6–15, scaling up the volumes as appropriate.

[a] Based on ref. 63.

Several splicing factors have been purified from Dignam-type extracts (29), which are highly active in splicing (26). Although it is logical to start with an extract that is active in the complete splicing reaction, it may be better to adopt a different procedure from that described in *Protocol 6* to obtain higher yields of factors, or to start with a more enriched subcellular fraction. An extract prepared from sonicated nuclei (see *Protocol 11*), although less active in splicing than a standard Dignam extract, appears to contain higher amounts of several activities necessary for splicing. These include SF2 and SC35. Its reduced overall splicing activity appears to be due to a high concentration of non-specific proteins and nucleic acids. The procedure is very similar to the large-scale preparation of standard nuclear extract (see *Protocol 10*), except that, after resuspending the isolated nuclei, sonication is carried out to disrupt the nuclear membrane and thus allow a more extensive extraction of nuclear contents.

Protocol 11. Preparation of nuclear extracts using sonication[a]

Equipment and reagents

- Frozen HeLa cells as described in *Protocol 10*, steps 1 and 2
- Branson model 450 sonicator with 0.5 inch diameter tip (or equivalent)
- Phase-contrast microscope
- Buffer C (see *Protocol 6*)
- Buffer C' (same as buffer C but with half as much KCl)
- All other reagents and materials described in *Protocol 6* and *Protocol 10*
- Polypropylene beaker (50 ml or size appropriate to the scale of the preparation)
- Rotating shaker

Method

1. Prepare a cell suspension as described in *Protocol 10*, steps 3–7.
2. Prepare a nuclear suspension in buffer C as described in *Protocol 6*, steps 6–10, scaling up the volumes as appropriate.

Continued

3: Splicing mRNA precursors

Protocol 11. *Continued*

3. Transfer the nuclear suspension to a polypropylene beaker, which should be two-thirds to three-quarters full.
4. Disrupt the nuclei by sonication. Place the beaker in a shallow ice-water bath and, with the power off, immerse the sonicator tip in the suspension to within 5 mm of the bottom, to avoid foaming. Use five pulses of 50% duration for 30 sec each, at an output of 100 watts or less. Wait at least 1 min between pulses to minimize any rise in temperature.
5. Monitor the sonication by phase-contrast microscopy and sonicate further, if necessary, to ensure that no nuclei remain intact.
6. Transfer the preparation to a centrifuge tube, place the tube on a rotating shaker for 30 min at 4 °C, and centrifuge as in *Protocol 6*, step 12.
7. Transfer the supernatant to a disposable polypropylene tube and keep it on ice.
8. Resuspend the pellet in an equal volume of buffer C', centrifuge as in step 6, and pool the supernatant with that from step 7. Discard the pellet containing chromatin fragments and nucleoli.
9. Finish preparing the extract as in *Protocol 6*, steps 13–15.

[a] Based on ref. 63.

4.1.3 Functional assays for monitoring purification of splicing factors

In most cases, the purification of a splicing factor is based on a functional assay. Sometimes it is possible to purify a factor by following an antigenic determinant for which antibodies are already available, or by electrophoretic analysis of an abundant protein or RNA constituent. However, factors may become inactivated during purification, and eventually it is desirable to demonstrate that the purified component is necessary for splicing.

The best functional assays consist of biochemical complementation assays in which a splicing extract is depleted of the specific factor under investigation. Addition of the partially- or highly-purified factor is then necessary for restoring splicing activity. The RNA products are analysed by urea–PAGE and autoradiography (Chapter 1, *Protocol 8* or Chapter 4, *Protocol 11*). Both of the RNA *trans*-esterification reactions involved in mRNA splicing are detectable, since the precursor (RNA substrate), intermediates, and products are electrophoretically distinguishable. Since each complementation assay has to be designed for the particular splicing factor under investigation, it is not feasible or appropriate to provide individual protocols here. Assays for specific factors are mentioned in the following sections. Further examples and general advice will be found in Chapters 4 and 5. A disadvantage of this approach is that the assay is laborious and time-consuming. One or two days may be necessary to dialyse the fractions, set up multiple complementation reactions, extract the RNA, run the gels, and obtain an autoradiogram. To avoid loss of activity during this time, the fractions may be stored at −70 °C. Sometimes it is preferable to dialyse aliquots, and only process the bulk of each fraction once it is clear which fractions are to be processed further and the next purification step has been chosen.

In addition to using activity assays, it is also helpful to monitor the progress of the purification by analysis of the protein and nucleic acid composition of the active fractions. The most useful methods for this are SDS–PAGE, with Coomassie blue or silver staining of proteins (see Chapter 5, *Protocols 7 and 8*), and urea–PAGE, with ethidium bromide staining of nucleic acids (see Chapter 1, *Protocol 8* or Chapter 4, *Protocol 11*). With this information, the active fractions can be pooled in such a way as to avoid unwanted protein or nucleic acid contaminants. Thus it may be wise to discard a subset of the active fractions if they overlap with unwanted proteins or RNAs.

Other assays that detect the products of splicing (such as primer extension to detect branched RNA) can be used in principle, although they require long times and multiple manipulations (65) (see also Chapter 2, Section 4). Factors can also be purified on the basis of the formation of specific complexes with appropriate RNAs, and in some cases, with other purified factors. Commonly-used assays in these cases include binding to nitrocellulose filters (66), UV cross-linking (67), and gel-retardation (68) (see also Chapter 5, Section 7). Radiolabelled RNA complexed with protein is retained on a nitrocellulose filter, forms a covalent protein–RNA adduct, or exhibits an altered electrophoretic mobility on a native gel (see Chapter 1, *Protocol 6*, or Chapter 4, *Protocol 12*). These assays are more rapid than full splicing assays, but additional work is necessary to demonstrate that the complexes are specific and necessary for splicing. This may require complete purification of the factor and the generation of antibodies.

A factor that is required for splicing in a biochemical complementation assay is not necessarily a splicing factor *per se*. For example, creatine phosphokinase is required for regenerating ATP in crude extracts or certain partially-purified fractions (26, 69). Similarly, in theory, a protease inhibitor could be purified as a putative splicing factor using complementation assays. Characterization of the fractions at an early stage may be sufficient to eliminate these and other trivial possibilities but, ultimately, detailed functional characterization of a purified factor is required to demonstrate a direct role in splicing.

One additional point deserves mention here. Occasionally, endogenous RNase inhibitors appear to be removed or inactivated in some fractions during purification, resulting in the activation of endogenous nucleases. These degrade the RNA substrate even in complementation assays, although RNA assembled in functional spliceosomes appears to be more resistant. Commercial RNase inhibitors (such as RNAsin) may be helpful, depending upon the nature of the endogenous nuclease. Although fractions that completely degrade the RNA substrate cannot be assayed in a meaningful way, partial degradation of the substrate can sometimes be tolerated by running the reaction products on high-percentage polyacrylamide gels, such that lariat products migrate more slowly than the precursor and its degradation products (65), which would otherwise prevent product detection by autoradiography. This procedure also allows detection of splicing intermediates and products even with relatively inefficient splicing, since it tolerates long autoradiographic exposures.

4.1.4 Devising a purification scheme

Purification schemes may be based on standard protein purification techniques. Most of the usual considerations apply when using these methods and in deciding the order of their use. Standard laboratory manuals should be consulted as necessary. However, the following points should be borne in mind:

(a) The fractionation of extracts may require alterations in the composition of Dignam buffer D in which the extracts are normally made and stored (see *Protocols 6, 10, 11*). For chromatography the salt concentration may need to be varied; NaCl and KCl are used most commonly, the advantage of the former being its higher solubility. Ammonium sulphate can be substituted for fractionation by salting-out and for hydrophobic interaction chromatography. Hydroxyapatite chromatography is commonly performed in phosphate buffers and EDTA should not be present. The pH can be varied when appropriate for chromatographic binding or elution. If proteolysis is not a problem, PMSF can be omitted; if it is, additional protease inhibitors can be used. Magnesium salts, urea, guanidinium chloride, ethylene glycol, and modified nucleosides may be required in special circumstances.

(b) Nucleic acids may need to be removed at an early stage. It is essential to obtain fractions that are as concentrated as possible, and, therefore, it is desirable to load columns at or near their saturation capacities. Unfortunately, most columns have limited capacities for crude extracts, especially when they contain large amounts of nucleic acids. In addition, nucleic acids interfere with ion-exchange chromatography, and it is common to obtain multiple peaks of activity, representing free and nucleic-acid bound factors. Of the numerous procedures commonly used to remove nucleic acids (e.g. RNase and DNase treatment), few are appropriate for subsequent reactions with pre-mRNA substrates. In addition, complete removal of the nucleic acids is not appropriate if one is also fractionating snRNPs. One useful procedure is to load the extract on to an anion exchanger such as DEAE–cellulose, which binds nucleic acids more strongly than proteins. At 0.1 M NaCl and pH 8.0, the flow-through will contain only proteins and no RNA or DNA, provided the column is not overloaded. NaCl (0.15 M) elutes additional proteins and a few RNA-containing particles, such as U1 snRNP. Higher salt concentrations elute additional snRNPs and other nucleic acids. Another useful procedure is to fractionate crude extracts on

CsCl gradients. Free RNA forms a pellet, protein remains at the top, and RNP particles that are not dissociated in high salt have intermediate buoyant densities. This procedure was first used for the partial purification of snRNPs (70), which were later shown to be active in splicing (71). Ultracentrifugation has also been used to remove nucleic acids in the form of RNPs (71, 72). However, many protein factors are initially present in large aggregates and may, therefore, be lost. Pre-incubation of extracts with ATP may dissociate large RNP aggregates, thus releasing soluble factors (73).

(c) Gel filtration chromatography is a useful method but a drawback is that it causes significant dilution of the purified fractions and this is often accompanied by loss of activity (see Section 4.1.1). Gel filtration can also be used instead of dialysis to remove small molecules and for buffer exchange prior to assay of the fractions. In this case, smaller columns can and should be used to avoid sample dilution.

(d) Affinity chromatography can also be useful, particularly employing immobilized synthetic nucleic acids (e.g. poly A, C, G, or U agaroses) or even RNA with specific binding sites, when these are known (67, 74, 75). Immobilized ATP may also be useful for purifying ATP-binding factors (76). Elution can normally be obtained by increasing the salt concentration or, in some cases, affinity elution can be achieved by competition with free ligand. Immunoaffinity chromatography is a powerful purification method but requires antibodies specific for the component of interest. It may also be difficult to elute the bound factor while retaining activity, although certain antibody–antigen interactions are sensitive to glycerol (77). Changes in the pH or the use of denaturants are more common elution methods (78), but these may result in the loss of activity. Antibodies to the trimethylguanosine 'cap' have been used to purify snRNPs, which can be gently eluted by competition with free 'cap' analogue (79, 80). Similarly, antibodies to peptides may be dissociated from bound protein antigens by competition with soluble peptide (81). High-performance FPLC columns are very useful for the last stages of a purification. The most common include Mono Q (anion exchanger), Mono S (cation exchanger), Phenyl Superose (hydrophobic interaction) and Superose 6 and 12 (gel filtration resins). For further details of immunoaffinity purification and ion-exchange chromatography of snRNPs see Chapter 5 (Sections 2 and 3).

(e) Inclusion of DTT and glycerol in the buffers may stabilize the splicing activity. Very low protein concentrations are usually detrimental to recovery and stability. In the final purification stages, carrier proteins such as nuclease-free BSA (0.1–1 mg/ml) may be included for optimal preservation of activity, but this should be avoided if the sample is to be subjected to amino acid composition or sequence analysis.

4.2 Purification procedures for specific splicing factors

4.2.1 Introduction
The following sections survey procedures for several factors that have been extensively purified. These do not include the numerous studies on general and specific RNA-binding proteins, including core and non-core hnRNP proteins, some of which are capable of binding to specific conserved splicing elements and/or related sequences, either in crude extracts or in purified form. Many of these factors are present in spliceosomes, and there are indications that some of these factors may be involved in splicing (reviewed in refs 60, 61).

4.2.2 Purification of multiple splicing activities from the same extract
Several specific procedures have been described for the stepwise separation and enrichment of multiple splicing activities simultaneously from a single extract. This is an arduous task which has not yet yielded homogeneously purified components. However, it has been a very useful approach to determine the minimum number of activities involved, and to define the combinations of distinct activities that are sufficient to promote successive steps in spliceosome assembly and RNA cleavage–ligation catalysis. The most extensive use of this fractionation/reconstitution strategy is described in refs 55, 82–85.

4.2.3 Purification of snRNPs
Several procedures have been developed for the isolation of U snRNPs, which yield particles with different degrees of purification. The purification of snRNPs can be readily followed by analysing the snRNAs by urea–PAGE (see Chapter 1, *Protocol 8* and Chapter 4, *Protocol 11*). The easiest functional assay is complementation of a micrococcal nuclease-treated extract (55) (see also Chapter 5, *Protocol 15*). Individual snRNPs may also be assayed after targeted digestion with RNase H and degradation of the specific oligonucleotide with DNase I (see Section 3.2.4 and Chapter 2, *Protocol 4* and Chapter 4, *Protocol 10*). However there is no guarantee that the snRNPs purified by complementation are structurally intact and fully active, as opposed to carrying partial activities.

A powerful method of purifying all the snRNPs is by immunoaffinity chromatography of crude extract with anti-trimethylguanosine monoclonal antibodies (79, 80). Several methods are available for covalently immobilizing the antibody (see Chaper 5, Section 2.2). Complete removal of snRNPs from the flowthrough has not been achieved, but it is recommended to use low flow rates, and/or recycling in order to maximize binding. *Protocol 12* describes the immunoaffinity purification of snRNPs using a monoclonal antibody against 2,2,7-trimethylguanosine (see also Chapter 5, *Protocol 3*).

Protocol 12. Immunoaffinity purification of snRNPs[a]

Equipment and reagents

- Monoclonal antibody against 2,2,7-trimethylguanosine bound to agarose as described in Chapter 5, *Protocol 2* (available from Oncogene Science, Inc.)
- HeLa cell nuclear extract (see *Protocol 10*)
- Buffer D (see *Protocol 6*)
- Buffer D with 20 mM 7-methylguanosine (Sigma)
- Buffer D with 6.0 M urea, made fresh, and lacking DTT
- Buffer D with 0.02% sodium azide
- Materials for denaturing polyacrylamide gel electrophoresis (Chapter 1, *Protocol 8* or Chapter 4, *Protocol 11*)
- Centriprep-30 ultrafiltration units (Amicon)

Method

1. Equilibrate the antibody–agarose in a chromatography column with buffer D[b]. Use 1 ml of matrix for up to 50 ml of nuclear extract.
2. Load the nuclear extract on to the column at 2 ml/h, or circulate it through the column at a faster flow rate for 16–24 h.
3. Wash the column with at least 20 column vol. of buffer D.
4. Elute the snRNPs in buffer D containing 20 mM 7-methylguanosine. Use a flow rate of 2.4 ml/h to minimize dilution of the snRNP fractions. Collect a minimum of 50 fractions (1 ml each).
5. Assay the fractions for snRNAs by phenol extraction, ethanol precipitation, and denaturing PAGE (Chapter 1, *Protocol 8* or Chapter 4, *Protocol 11*).
6. Regenerate the column by washing it in 2 vol. of buffer D lacking DTT followed by 2 vol. of buffer D with 6 M urea and lacking DTT, and finally 10 vol. of buffer D with 0.02% sodium azide as a preservative.
7. Concentrate the snRNPs by ultrafiltration using the Centriprep-30 ultrafiltration units using the method recommended by the manufacturer.

[a] Based on refs 79, 80.
[b] The buffer may contain 0.1–0.3 M KCl or NaCl, and 5% glycerol.

Specific removal of individual snRNPs by affinity chromatography with 2′-O-methyloligoribonucleotides is another way of generating complementing fractions (58, 59) (see Chapter 4, *Protocol 7*). The use of modified oligonucleotides serves primarily to protect the snRNA against endogenous RNase H, which would otherwise degrade the snRNA (54, 55). However, despite this precaution, the depleted extract may still contain free snRNP proteins, which are probably present in excess. It should be possible to recover highly purified particles from the oligo (2′-O-alkylribonucleotide) column, for example by lowering the salt concentration, raising the temperature, or otherwise destabilizing base-pairing.

Fractionation on CsCl gradients is an alternative, simple procedure that takes advantage of the buoyant densities of snRNPs, which are intermediate between protein and RNA. In the presence of magnesium ions, the particles appear to be stabilized so that they are not dissociated in the presence of molar amounts of CsCl (70). The purification achieved is not as extensive as in the case of immunoaffinity chromatography.

Further separation of snRNPs into individual classes of particles is difficult, in part because of their similar composition. Although it is possible to enrich individual particles by various forms of chromatography (86), it is common to observe heterogeneous chromatographic behaviour. For example, multiple peaks of U1 have been observed due to the loss of some polypeptides, either prior to or during chromatography (87). The existence of multi-snRNP complexes, in addition to free snRNPs, also causes chromatographic heterogeneity.

Although the roles of the majority of snRNP polypeptides in splicing have not been established, biochemical experiments are beginning to identify specific functions for some of these polypeptides. For example, the U1-specific C protein appears to stimulate binding of the U1 snRNP to 5' splice sites (87). One or more proteins associated with U5 snRNP may direct it to bind to 3' splice sites (40, 88), and these or other U5 proteins are required for assembly of a U4/U6/U5 multi-snRNP complex (89). Binding of this snRNP complex to the assembling spliceosome is disrupted in extracts prepared from heat-shocked cells, and can be restored by the U5-associated proteins (89).

4.2.4 Purification of SF2/ASF

SF2 is a human splicing factor originally identified in a complementation assay for the first *trans*-esterification reaction of splicing (55, 63). This factor, also known as ASF (37), was subsequently shown to be an RNA-binding protein required for early spliceosome assembly steps and capable of modulating the selection of alternative 5' splice sites in a concentration-dependent manner (37, 63, 90).

Unlike other splicing factors, SF2 is retained within the nuclei when the hypotonic lysis method is used to prepare splicing extracts (see *Protocol 6*, refs. 26, 29). Thus the postnuclear (S100) extract is SF2-dependent and can form the basis of a simple complementation assay for monitoring SF2 purification (63). *Protocol 13* describes the preparation of postnuclear S100 extract by hypotonic lysis, which can be carried out in conjunction with the standard nuclear extract preparation (see *Protocol 6*).

Protocol 13. Preparation of a HeLa cell postnuclear S100 extract by hypotonic lysis[a]

Equipment and reagents

- Buffer B (0.3 M Hepes–KOH, pH 7.9, 1.4 M KCl, 30 mM $MgCl_2$)
- Buffer D (as in *Protocol 6*)

Continued

Protocol 13. *Continued*

- All other reagents, cells, and materials as described in *Protocol 6*
- Ultracentrifuge with fixed angle rotor
- (Beckman Type 70 Ti or equivalent) and polycarbonate ultracentrifuge bottles

Method

1. Prepare a cell homogenate and pellet the nuclei as described in *Protocol 6*, steps 1–7.
2. Transfer the supernatant to a polypropylene measuring cylinder and measure the volume.
3. Add 0.11 vol. of buffer B and mix.
4. Transfer the extract to ultracentrifuge polycarbonate bottles.
5. Centrifuge at 100 000 g_{av} for 1 h at 4 °C.
6. Dialyse the S100 supernatant against buffer D and finish preparing the postnuclear extract as described for the nuclear extract in *Protocol 6*, steps 13–15.

[a] Based on ref. 29.

Protocol 14 describes the purification of SF2 from a nucleoplasmic extract prepared by sonication (see *Protocol 11*). Precipitation with 50–80% ammonium sulphate recovers all of the SF2 but only 10% of the total protein, and even less nucleic acid. This substantial degree of purification makes it possible to concentrate the SF2 fraction. Occasionally, SF2 will precipitate after dialysis; the dialysed suspension retains activity, but is not suitable for further column chromatography. To purify SF2 further by column chromatography, dialysis of the ammonium sulphate fraction should be omitted.

Protocol 14. Purification of SF2/ASF[a]

Equipment and reagents

- HeLa cell nuclear extract prepared by sonication (see *Protocol 11*)
- Buffer E (20 mM Hepes–KOH, pH 8.0, 0.2 mM EDTA, 1 mM DTT)
- Saturated ammonium sulphate solution in buffer E (add 69.7 g salt to 100 ml of buffer E)
- Buffer F (same as buffer D in *Protocol 6*, except with 5% glycerol)
- Buffer G (same as buffer F except with 3.4 M ammonium sulphate instead of 0.1 M KCl)
- Buffer H (same as buffer G except with 2.0 M ammonium sulphate)
- Buffer I (same as buffer F except lacking salt)
- Sephacryl S200HR (100 × 1.6 cm), Mono Q (1 ml) and Phenyl Superose (1 ml) columns (Pharmacia–LKB)
- Materials for splicing reactions (see *Protocol 7*)
- Postnuclear (S100) extract (see *Protocol 13*)

Continued

3: *Splicing mRNA precursors*

Protocol 14. *Continued*

Method

1. Add 1.5 vol. of buffer E to the HeLa cell nuclear extract.
2. Add an equal volume of saturated ammonium sulphate slowly with gentle stirring and continue stirring for 1 h at 4 °C.
3. Centrifuge the mixture at 4000–5000 g_{max} for 30 min at 4 °C.
4. Transfer the supernatant to a fresh container, measure the volume, and add 19.4 g of solid ammonium sulphate per 100 ml of supernatant.
5. Stir gently for 1 h and centrifuge the suspension as in step 3.
6. Aspirate the supernatant and redissolve the pellet in 0.1 vol. (relative to the starting extract, step 1) of buffer I.
7. Fractionate the material on a Sephacryl S200HR column in buffer F. For a 4 ml sample use a 200 ml column and collect 5 ml fractions at 0.5 ml/min.
8. Assay the fractions for SF2 using the HeLa cell postnuclear (S100) extract (see *Protocol 13*). Set up the splicing reactions and carry out the analysis as in *Protocol 7*, but replace the usual nuclear extract with 0.5 vol. of S100 extract and 0.5 vol. of an SF2 fraction.
9. Pool the SF2 peak fractions, usually representing the excluded volume of the Sephacryl column, since SF2 elutes as a complex with residual RNA.
10. Fractionate the pooled SF2 fractions on a Mono Q column, eluting the column with 5 ml steps of buffer F, buffer F with 0.35 M KCl, and buffer F with 1 M KCl. Use a flow rate of 0.5 ml/min and collect 0.5 ml fractions.
11. Dialyse the fractions against buffer F and then assay the fractions as in step 8. SF2 should be present in the last salt step.
12. Pool the SF2-containing fractions and add 1.5 vol. of buffer G.
13. Load the mixture on to a Phenyl Superose column in buffer H, eluting the column with 5 ml of buffer H and then with a 25 ml gradient from buffer H to buffer I. Use a flow rate of 0.4 ml/min and collect 1 ml fractions.
14. Dialyse the fractions (as in step 11) and assay them for SF2 (see step 8). SF2 emerges very early in the gradient.

a Based on ref. 63.

An alternative purification of SF2 from 293 cells has been published by Ge and Manley (37), based on the alternative splice site selection assay.

Further optimization of the initial ammonium sulphate fractionation procedure described in *Protocol 14* (steps 1–5) shows that active SF2 can be recovered in a 60% to 90% fraction, which can be redissolved in a small volume and dialysed. This fraction can then be adjusted to 20 mM $MgCl_2$ and, remarkably, SF2 and related activities quantitatively precipitate, for unknown reasons. The pellet can be rinsed with 20 mM $MgCl_2$ in buffer D, and redissolved in buffer D to yield a highly active and enriched preparation of

SF2 (91). If this procedure is performed with Dignam extract, the predominant polypeptide is SF2, although other less abundant polypeptides that also have SF2 activity are present as well.

A 28 kDa polypeptide is sufficient for activity in both the complementation and alternative splice site selection assays (92, 93). SF2/ASF cDNAs have been cloned and expressed in *E. coli* (92, 93). The recombinant protein lacks the phosphate modifications present in the human protein, and has an apparent M_r of 31.5 kDa. The modified protein migrates typically as a doublet of about 33 kDa. An additional polypeptide with an apparent M_r of 32 kDa copurifies with SF2 but is not required for complementation. It is unrelated to SF2 in sequence, and may or may not be a subunit (90, 93).

Recombinant SF2/ASF with correct N- and C-termini has different purification properties compared to the protein isolated from HeLa cells, and is either present in inclusion bodies or, if expressed at lower levels, it becomes insoluble during the purification. The recombinant protein can be solubilized in SDS, guanidinium chloride, or urea and purified under denaturing conditions, e.g. by preparative SDS–PAGE. SF2/ASF then has to be renatured by controlled removal of the denaturing agent, yielding reasonably active preparations (92, 93). Preparative denaturing electrophoresis and refolding have also been used to characterize SF2 and related polypeptides from *Drosophila melanogaster* (91).

4.2.5 Purification of SF5/hnRNP A1

SF5 was recently identified as an activity that counteracts the concentration-dependent effects of SF2 on alternative splicing (Section 4.2.4) (94). Whereas high concentrations of SF2 promote utilization of proximal alternative 5′ splice sites *in vitro*, high concentrations of SF5 favour distal alternative 5′ splice sites. It is not known whether this factor is also essential for splicing *per se*. The purification assay consisted of adding fractions and partially purified SF2 to a postnuclear S100 extract prepared by hypotonic lysis (see *Protocol 13*) and looking for a switch in 5′ splice site selection during splicing of a pre-mRNA containing alternative 5′ splice sites. For details of this assay procedure see ref. 94. The purification of SF5 is described in *Protocol 15*. SF5 was shown to be identical to hnRNP A1, a well-characterized core hnRNP protein.

Protocol 15. Purification of SF5/hnRNP A1[a]

Equipment and reagents

- HeLa cell nuclear extract prepared by sonication (see *Protocol 11*)
- Buffer F, buffer I, saturated ammonium sulphate solution and Phenyl Superose column (see *Protocol 14*)
- Buffer J (Buffer F containing CsCl to a final solution density of 1.3 g/ml). Add 4.5 g of CsCl to 10 ml of buffer F
- Buffer K (Buffer F containing 2 M NaCl instead of KCl)

Continued

Protocol 15. Continued

- Gradient fractionator (Buchler Auto-densiflow or equivalent)
- Ultracentrifuge with fixed angle rotor
- (Beckman Type 70 Ti or equivalent) and polycarbonate ultracentrifuge bottles

Method

1. Prepare an ammonium sulphate fraction of the nuclear extract as described in *Protocol 14*, steps 1–3. SF5 is in the pellet and SF2 is in the supernatant.
2. Redissolve the pellet in buffer F (same vol. as the starting extract).
3. Dialyse the redissolved pellet against buffer F and remove insoluble material by centrifugation at 25 000 g_{av} for 20 min at 4 °C.
4. Add 1 g of CsCl per ml of sample and allow it to dissolve (final solution density of 1.6 g/ml).
5. Half-fill the polycarbonate bottles with buffer J. Underlay with an equal vol. of the sample/CsCl solution.
6. Centrifuge the gradients at 160 000 g_{av} for 24 h at 3 °C.
7. Fractionate the gradients from the top, collecting 0.5 ml fractions.
8. Dialyse the fractions against buffer F and assay for SF5 (see text). Pool the peak fractions.
9. Dialyse the pooled SF5 fractions against buffer K and remove insoluble material by centrifugation for 10 min in a microcentrifuge.
10. Fractionate a portion of the sample (5 mg protein) on a Phenyl Superose column (1 ml) in buffer K, eluting the column with a 20 ml gradient from buffer K to buffer I. Use a flow rate of 0.3 ml/min and collect 1 ml fractions.
11. Dialyse and assay the fractions and pool the peak SF5 activity (as in step 8).

[a] Based on ref. 94.

SF5/hnRNP A1 has been overproduced in *E. coli* from cloned cDNA (94). The protein is expressed at high levels in soluble form and is as active in the splice site selection assay as the purified human protein. *Protocol 15* can be used for purifying the recombinant protein using a bacterial lysate instead of HeLa nuclear extract (94).

4.2.6 Purification of U2AF

This activity was originally described as an auxiliary activity required to obtain ATP-dependent protection of the branchpoint by fractions containing U2 snRNP (71). It was purified on the basis of this assay and shown to consist of two subunits of 65 kDa and 35 kDa (95). More recently, an activity assay was developed to show that the 65 kDa subunit is required for splicing (64).

Purification schemes for U2AF have recently been described (84, 95). Both take advantage of the extremely tight binding of U2AF to poly(U)-Sepharose. Recently, a cDNA for the 65 kDa subunit of U2AF was isolated. Using mRNA prepared from this cDNA, the subunit has been translated in wheatgerm extracts, with sufficient efficiency to complement U2AF-depleted extract (96).

4.2.7 Purification of SC35

This factor was originally defined as a spliceosome constituent by a specific monoclonal antibody that was raised against spliceosome-containing fractions obtained by gel filtration of a preparative splicing reaction (52). Nuclear extract immunodepleted of SC35 is inactive in the first *trans*-esterification reaction and immunoaffinity-purified SC35 can complement the immunodepleted extract (53). The antibody reacts with a 35 kDa doublet in Western blotting. Recently, a cDNA encoding the 35 kDa polypeptide was cloned and found to encode a polypeptide that bears some similarity to SF2/ASF (97).

4.2.8 Purification of an 88 kDa polypeptide (SF 53/4)

This factor was identified by its reactivity with monoclonal antibody 53/4 raised against a 200S fraction obtained from HeLa nuclear extract run on sucrose gradients. The fraction contained large RNPs (51). The antibody reacts with an 88 kDa polypeptide by Western blotting and immunoprecipitation, and can be used to purify this protein by immunoaffinity chromatography with elution at acidic pH. The immunodepleted extract is inactive in spliceosome assembly and splicing, and can be complemented with the purified 88 kDa protein (51).

4.2.9 Purification of HRF

Heat Reversal Factor (HRF) is an activity that can complement heat-inactivated nuclear extracts to restore splicing (76). Mild heat treatment (45 °C, 10 min) selectively inhibits the second *trans*-esterification step of splicing (55) and a slightly harsher heat treatment (46 °C for 20 min) also inhibits the first step, as well as spliceosome assembly (76). HRF can restore spliceosome assembly and the first reaction. Surprisingly, purified HRF is quite heat resistant (95 °C for 30 min), suggesting that the heat sensitivity of extracts is mediated through interactions of HRF with other components (76). Alternatively, the splicing factor might be heat sensitive and HRF is required for activating this labile component. Therefore, it is not clear at present whether HRF is a splicing factor or, for example, a heat-shock factor.

The reported purification (76), based on complementation of the heat-inactivated extract, involves chromatography of the crude nuclear extract (see *Protocol 10*) on DEAE-Sepharose CL6B, ATP-agarose and Phenyl-Sepharose columns, as well as a boiling step prior to the last column.

4.3 Isolation of cDNAs for splicing factors and expression of the recombinant proteins

In most cases, an important goal is to isolate a cDNA encoding the protein of interest both to determine its sequence and to allow overproduction of recombinant protein. If antibodies specific to the purified protein are available, they can be used to screen mammalian cDNA expression libraries (98). However, this approach may fail if the antibodies recognize modified or conformational epitopes that cannot be generated in *E. coli*. An alternative approach is to obtain partial amino acid sequence from the N-terminus of the purified protein, or from one or more internal peptides generated by chemical or enzymatic cleavage. The amino acid sequence can be used to design degenerate oligonucleotides to screen cDNA libraries by hybridization, or first to obtain unique DNA probes by RT-PCR. Section 2.2.3 of this chapter and Chapter 2 (Section 3.4) describe aspects of RT-PCR (see also ref. 99). Methods for cDNA cloning and for screening cDNA libraries by hybridization or with antibodies are described in other volumes of this Series (100-102). Examples of overproduction of active recombinant splicing factors in bacteria or baculovirus are given in refs 53, 92-94.

Acknowledgements

The authors are grateful to Dr B. Chabot for comments on the manuscript.

References

1. Pouwels, P. H., Enger-Valk, B. E., and Brammer, W. J. (1985 and supplements). *Cloning vectors: a laboratory manual*. Elsevier, Amsterdam.
2. Gluzman, Y. (1981). *Cell*, **23**, 175.
3. Libri, D., Marie, J., Brody, E., and Fiszman, M. Y. (1989). *Nucleic Acids Res.*, **17**, 6449.
4. Guo, W., Mulligan, G. J., Wormsley, S., and Helfman, D. M. (1991). *Genes Dev.*, **5**, 2096.
5. Huang, M. T. F. and Gorman, C. M. (1990). *Mol. Cell. Biol.*, **10**, 1805.
6. Spandidos, A. D. and Wilkie, N. M. (1984). In *Transcription and translation: a practical approach* (ed. B. D. Hames and S. J. Higgins), pp. 1-48. IRL Press, Oxford.
7. Kriegler, M. (1990). *Gene transfer and expression: a laboratory manual*. Stockton Press, New York.
8. Gallego, M. E. and Nadal-Ginard, B. (1990). *Mol. Cell. Biol.*, **10**, 2133.
9. Nasim, F. H., Spears, P. A., Hoffmann, H. M., Kuo, H., and Grabowski, P. J. (1990). *Genes Dev.*, **4**, 1172.
10. Crenshaw, E. B., Russo, A. F., Swanson, L. W., and Rosenfeld, M. G. (1987). *Cell*, **49**, 380.
11. Dillon, N. and Grosveld, F. (1993). In *Gene transcription: a practical approach* (ed. B. D. Hames and S. J. Higgins), p. 153. IRL Press, Oxford.
12. Sosnowski, B. A., Belote, J. M., and McKeown, M. (1989). *Cell*, **58**, 449.
13. Emerson, R. B., Hedjran, F., Yeakley, J. M., Guise, J. W., and Rosenfeld, M. G. (1989). *Nature*, **341**, 76.

14. Sisodia, S. S., Cleveland, D. W., and Sollner-Webb, B. (1987). *Nucleic Acids Res.*, **15**, 1995.
15. Clark, J. M. (1988). *Nucleic Acids Res.*, **16**, 9677.
16. Melton, D. A., Krieg, P. A., Rebagliati, M. R., Maniatis, T., Zinn, K., and Green, M. R. (1984). *Nucleic Acids Res.*, **12**, 7035.
17. Nakayama, H., Yokoi, H., and Fujita, J. (1992). *Nucleic Acids Res.*, **20**, 4939.
18. Wiesner, R. J. (1992). *Nucleic Acids Res.*, **20**, 5863.
19. Goblet, C., Prost, E., and Whalen, R. G. (1989). *Nucleic Acids Res.*, **17**, 2144.
20. Don, R. H., Cox, P. T., Wainwright, B. J., Baker, K., and Mattick, J. S. (1991). *Nucleic Acids Res.*, **19**, 4008.
21. Hamshere, M., Dickson, G., and Eperon, I. (1991). *Nucleic Acids Res.*, **19**, 4709.
22. Eperon, I. C. (1986). *Analyt. Biochem.*, **156**, 406.
23. Eperon, L. P., Graham, I. R., Griffiths, A. D., and Eperon, I. C. (1988). *Cell*, **54**, 393.
24. Olson, C. H., Blattner, F. R., and Daniels, D. L. (1991). *Methods: a companion to methods in enzymology*, **3**, 27.
25. Green, M. R., Maniatis, T., and Melton, D. A. (1983). *Cell*, **32**, 681.
26. Krainer, A. R., Maniatis, T., Ruskin, B., and Green, M. R. (1984). *Cell*, **36**, 993.
27. Hernandez, N. and Keller, W. (1983). *Cell*, **35**, 89.
28. Hardy, S. F., Grabowski, P. J., Padgett, R. A., and Sharp, P. A. (1984). *Nature*, **308**, 375.
29. Dignam, J. D., Lebovitz, R. M., and Roeder, R. G. (1983). *Nucleic Acids Res.*, **11**, 1475.
30. Kedes, D. H. and Steitz, J. A. (1987). *Proc. Natl Acad. Sci. USA*, **84**, 7928.
31. Noble, J. C. S., Pan, Z.-Q., Prives, C., and Manley, J. L. (1987). *Cell*, **50**, 227.
32. Cote, G. J., Nguyen, I. N., Lips, J. M. C., Berget, S. M., and Gagel, R. F. (1991). *Nucleic Acids Res.*, **19**, 3601.
33. Lee, K. A. W., Bindereif, A., and Green, M. R. (1988). *Gene Anal. Techn.*, **5**, 22.
34. La Branche, H., Frappier, D., and Chabot, B. (1991). *Nucleic Acids Res.*, **19**, 4509.
35. Rio, D. C. (1988). *Proc. Natl Acad. Sci. USA*, **85**, 2904.
36. Hannon, G. J., Maroney, P. A., Denker, J. A., and Nilsen, T. W. (1990). *Cell*, **61**, 1247.
37. Ge, H. and Manley, J. L. (1990). *Cell*, **62**, 25.
38. Siebel, C. W. and Rio, D. C. (1990). *Science*, **248**, 1200.
39. Black, D. L. (1992). *Cell*, **69**, 795.
40. Tazi, J., Alibert, C., Temsamani, J., Reveillaud, I., Cathala, G., Brunel, C., and Jeanteur, P. (1986). *Cell*, **47**, 755.
41. Murthy, K. G. K., Park, P., and Manley, J. L. (1991). *Nucleic Acids Res.*, **19**, 2685.
42. Kreivi, J.-P., Zerivitz, K., and Akusjärvi, G. (1992). *Nucleic Acids Res.*, **19**, 6956.
43. Reed, R. and Maniatis, T. (1986). *Cell*, **46**, 681.
44. Mayeda, A. and Ohshima, Y. (1988). *Mol. Cell. Biol.*, **8**, 4484.
45. Cunningham, S. A., Else, A. J., Potter, B. V. L., and Eperon, I. C. (1991). *J. Mol. Biol.*, **217**, 265.
46. Schmitt, P., Gattoni, R., Keohavong, P., and Stévenin, J. (1987). *Cell*, **50**, 31.
47. Norton, P. A. and Hynes, R. O. (1990). *Nucleic Acids Res.*, **18**, 4089.
48. Ruskin, B. and Green, M. R. (1985). *Cell*, **43**, 131.
49. Rymond, B. C. and Rosbash, M. (1986). *EMBO J*, **5**, 3517.
50. Krämer, A., Keller, W., Appel, B., and Lührmann, R. (1984). *Cell*, **38**, 299.

51. Ast, G., Goldblatt, D., Offen, D., Sperling, J., and Sperling, R. (1991). *EMBO J.*, **10**, 425.
52. Fu, X.-D. and Maniatis, T. (1990). *Nature*, **343**, 437.
53. Spector, D. L., Fu, X.-D., and Maniatis, T. (1991). *EMBO J.*, **10**, 3467.
54. Black, D. L., Chabot, B., and Steitz, J. A. (1985). *Cell*, **42**, 737.
55. Krainer, A. R. and Maniatis, T. (1985). *Cell*, **42**, 725.
56. Pan, Z.-Q. and Prives, C. (1989). *Genes Dev.*, **3**, 1887.
57. Hamm, J., Dathan, N. A., and Mattaj, I. W. (1989). *Cell*, **59**, 159.
58. Barabino, S. M. L., Blencowe, B. J., Ryder, U., Sproat, B. S., and Lamond, A. I. (1990). *Cell*, **63**, 293.
59. Lamm, G. M., Blencowe, B. J., Sproat, B. S., Iribarren, A. M., Ryder, U., and Lamond, A. I. (1991). *Nucleic Acids Res.*, **19**, 3193.
60. Krainer, A. R. and Maniatis, T. (1988). In *Transcription and splicing* (ed. B. D. Hames and D. M. Glover), pp. 131–206. IRL Press, Oxford.
61. Green, M. R. (1991). *Annu. Rev. Cell Biol.*, **7**, 559.
62. Ruby, S. W. and Abelson, J. (1991). *Trends Genet.*, **7**, 79.
63. Krainer, A. R., Conway, G. C., and Kozak, D. (1990). *Genes Dev.*, **4**, 1158.
64. Zamore, P. D. and Green, M. R. (1991). *EMBO J.*, **10**, 207.
65. Ruskin, B., Krainer, A. R., Maniatis, T., and Green, M. R. (1984). *Cell*, **38**, 317.
66. Mayeda, A., Tatei, K., Kitayama, H., Takemura, K., and Oshima, Y. (1986). *Nucleic Acids Res.*, **14**, 3045.
67. García-Blanco, M. A., Jamison, S. F., and Sharp, P. A. (1989). *Genes Dev.*, **3**, 1874.
68. Konarska, M. M. and Sharp, P. A. (1986). *Cell*, **46**, 845.
69. Krämer, A. and Keller, W. (1990). *Methods in enzymology*, Vol. 181 (ed. J. E. Dahlberg and J. N. Abelson) p. 3. Academic Press, San Diego.
70. Lelay-Taha, M.-N., Reveillaud, I., Sri-Widada, J., Brunel, C., and Jeanteur, P. (1986). *J. Mol. Biol.*, **189**, 519.
71. Ruskin, B., Zamore, P. D., and Green, M. R. (1988). *Cell*, **52**, 207.
72. Zapp, M. L. and Berget, S. M. (1989). *Nucleic Acids Res.*, **17**, 2655.
73. Conway, G. C., Krainer, A. R., Spector, D. L., and Roberts, R. J. (1989). *Mol. Cell. Biol.*, **9**, 5273.
74. Grabowski, P. J. and Sharp, P. A. (1986). *Science*, **233**, 1294.
75. Reed, R. (1990). *Proc. Natl Acad. Sci. USA*, **87**, 8031.
76. DeLannoy, P. and Caruthers, M. H. (1991). *Mol. Cell. Biol.*, **11**, 3425.
77. Thompson, N. E., Aronson, D. B., and Burgess, R. R. (1990). *J. Biol. Chem.*, **265**, 7069.
78. Harlow, E. and Lane, D. (1988). In *Antibodies: a laboratory manual*, pp. 471–510. Cold Spring Harbor Laboratory, Cold Spring Harbor, NY.
79. Bochnig, P., Reuter, R., Bringmann, P., and Lührmann, R. (1987). *Eur. J. Biochem.*, **168**, 461.
80. Krainer, A. R. (1988). *Nucleic Acids Res.*, **16**, 9415.
81. Behrens, S. E. and Lührmann, R. (1991). *Genes Dev.*, **5**, 1439.
82. Krämer, A., Frick, M., and Keller, W. (1987). *J. Biol. Chem.*, **262**, 17630.
83. Krämer, A. (1988). *Genes Dev.*, **2**, 1155.
84. Krämer, A. and Utans, U. (1991). *EMBO J.*, **10**, 1503.
85. Perkins, K. K., Furneaux, H. M., and Hurwitz, J. (1986). *Proc. Natl Acad. Sci. USA*, **83**, 887.

86. Krämer, A. (1990). *Methods in enzymology*, Vol. 181, (ed. J. E. Dahlberg and J. N. Abelson), p. 215. Academic Press, San Diego.
87. Heinrichs, V., Bach, M., Winkelmann, G., and Lührmann, R. (1990). *Science*, **247**, 69.
88. Gerke, V. and Steitz, J. A. (1986). *Cell*, **47**, 973.
89. Utans, U., Behrens, S.-E., Lührmann, R., Kole, R., and Krämer, A. (1992). *Genes Dev.*, **6**, 631.
90. Krainer, A. R., Conway, G. C., and Kozak, D. (1990). *Cell*, **62**, 35.
91. Mayeda, A., Zahler, A. M., Krainer, A. R., and Roth, M. B. (1992). *Proc. Natl Acad. Sci. USA*, **89**, 1301.
92. Ge, H., Zuo, P., and Manley, J. L. (1991). *Cell*, **66**, 373.
93. Krainer, A. R., Mayeda, A., Kozak, D., and Binns, G. (1991). *Cell*, **66**, 383.
94. Mayeda, A. and Krainer, A. R. (1992). *Cell*, **68**, 365.
95. Zamore, P. D. and Green, M. R. (1989). *Proc. Natl Acad. Sci. USA*, **86**, 9243.
96. Zamore, P. D., Patton, J. G., and Green, M. R. (1992). *Nature*, **355**, 609.
97. Fu, X.-D. and Maniatis, T. (1992). *Science*, **256**, 535.
98. Young, R. A. and Davis, R. W. (1983). *Proc. Natl Acad. Sci. USA*, **80**, 1194.
99. McPherson, M. J., Quirke, P., and Taylor, G. M. (1991). *PCR: a practical approach*. IRL Press, Oxford.
100. Hames, B. D. and Higgins, S. J. (1985). *Nucleic acid hybridization: a practical approach*. IRL Press, Oxford.
101. Glover, D. M. (1985). *DNA cloning: a practical approach*, Volume I, IRL Press.
102. Brown, T. A. (1991). *Essential molecular biology: a practical approach*, Volume I, IRL Press, Oxford.
103. Eperon, I. C., Ireland, D. C., Smith, R. A., Mayeda, A., and Krainer, A. R. (1993). *EMBO J.*, in press.

4

Isolation and characterization of ribonucleoprotein complexes

ANGUS I. LAMOND and BRIAN S. SPROAT

1. Introduction

Detailed analyses of a wide variety of RNA processing reactions *in vitro* have shown that the processing machinery usually either comprises ribonucleoprotein (RNP) complexes or is dependent upon the formation of RNP complexes during the reaction mechanism. Major experimental goals for the analysis of RNA processing mechanisms therefore include:

- isolation of RNP complexes from cell extracts
- specific depletion of RNP complexes from cell extracts
- analysis of proteins associated with RNP complexes
- functional inactivation of RNP complexes

This chapter describes methods that allow the rapid and efficient isolation, depletion, or characterization of individual RNP complexes. In particular, it focuses upon recently developed techniques which rely upon the use of antisense oligonucleotides made of either DNA, RNA, or 2'-O-alkyl RNA (NB alkyl is the generic term used to represent the family of non-aromatic carbon chains, in this context usually methyl or allyl—see *Figure 1*). In these techniques it is the RNA components of individual RNP complexes which are targeted through the use of complementary antisense probes. The major alternative approach involves the use of antibodies to target the proteins present in individual RNP particles. Immunological techniques are presented in Chapter 5.

As most of the techniques discussed in this chapter involve the synthesis of oligonucleotides for use as antisense probes, it starts by describing current protocols for chemically synthesizing RNA and 2'-O-alkyl RNA.

2. Chemical synthesis of oligoribonucleotides and oligo(2′-O-alkylribonucleotides)

2.1 Advantages of using antisense oligonucleotides for studying RNP structure and function

The chemical synthesis of oligoribonucleotides is of increasing importance to the study of RNA biochemistry and RNA processing as it enables the production of relatively large quantities of highly defined substrates and oligonucleotides with nucleotide sequences tailor-made to target any RNA or RNP involved in processing reactions. Using chemical synthesis it is possible to place modified bases and/or sugars at any given nucleotide in an RNA. Useful reporter groups can be incorporated at sequence-specific positions, for example psoralen for cross-linking purposes or biotin for affinity-based purification or depletion procedures (see Section 3). This control over the precise structure of RNA substrates cannot be achieved using the widely employed *in vitro* transcription systems based on phage RNA polymerases (see Chapter 1, Section 2.2). Chemically-substituted oligoribonucleotides, especially oligo(2′-O-alkylribonucleotides), are characterized by their resistance to both chemical and nuclease degradation. In addition, these substituted oligoribonucleotides form particularly stable hybrids with the target RNA sequence. This often allows somewhat shorter oligonucleotides and/or lower hybridization temperatures to be employed. RNA sequences predicted to form double-stranded stem structures can sometimes be targeted successfully, since the antisense probe may be able to displace the intramolecular secondary structure, again by forming a more stable hybrid with its complementary strand in the stem.

Due to the ease with which oligonucleotides can now be synthesized, the antisense approach can have advantages over immunological approaches. However, it should be stressed that ideally both immunological and antisense approaches should be used in parallel to characterize RNP structure and function.

2.2 General points concerning the chemical synthesis of oligoribonucleotides

Current protocols for the chemical synthesis of oligoribonucleotides, oligo(2′-O-methylribonucleotides), and oligo(2′-O-allylribonucleotides) will be given in the following sections. Their synthesis uses similar principles, chemistry, and instrumentation to those involved in oligodeoxyribonucleotide synthesis. The disadvantage with chemical synthesis is that its application remains limited by present-day technology to the production of relatively short RNA chains of up to about 60 nt in length.

Progress in the solid-phase chemical synthesis of oligoribonucleotides has been hampered due to the difficulty in developing appropriate orthogonal protecting groups; however, in the past few years, suitable 2′-hydroxyl protecting groups have been developed. Two such groups are the trialkylsilyl group, in particular tert.-butyldimethylsilyl (1, 2) and the 1-(2-fluorophenyl)-4-methoxypiperidin-4-yl (Fpmp) group (3). Suitable nucleotide monomers, 5′-O-dimethoxytrityl-2′-O-trialkylsilylribonucleoside-3′-O-(2-cyanoethyl N,N-diisopropylphosphoramidites) and functionalized controlled pore glass supports, for use in solid-phase synthesis by the 2-cyanoethyl phosphoramidite method, are commercially available from Milligen (benzoyl protection is used for the exocyclic amino groups of A, C, and G; the trialkylsilyl group used is the tert.-butyldimethylsilyl group for A, C, and U residues, and the triisopropylsilyl group for G residues). 2′-O-Fpmp protected monomers are available from Cruachem.

Figure 1 shows the general structure of a chemically synthesized oligoribonucleotide bound to its solid support and with its functional groups fully protected in the manner described above.

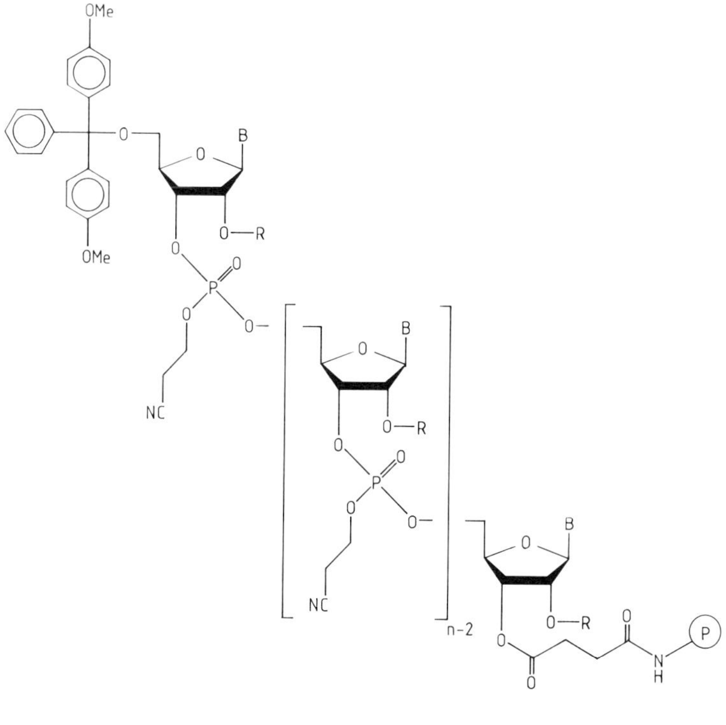

Figure 1. Structure of a carrier-bound, fully-protected synthetic oligoribonucleotide or oligo(2′-O-alkylribonucleotide) n residues long, prepared by the 2-cyanoethylphosphoramidite method. B, the base, may be the uracil-1-yl, N^4-acylcytosin-1-yl, N^2-acylguanin-9-yl, N^6-acyladenin-9-yl, hypoxanthin-9-yl or N^2,N^6-*bis*(phenoxyacetyl)-2,6-diaminopurin-9-yl group depending on the sequence desired; the acyl group is usually the benzoyl group. P represents the solid-phase support. R is a trialkylsilyl group in the case of an oligoribonucleotide (tert. butyldimethylsilyl for A, C, and U nucleotides, triisopropylsilyl for G nucleotides), a methyl group in the case of an oligo(2′-O-methylribonucleotide) or an allyl group in an oligo(2′-O-allylribonucleotide).

Assembly of the fully-protected carrier-bound oligoribonucleotide is carried out by a method directly analogous to that for an oligodeoxyribonucleotide, except that the coupling time using tetrazole has to be extended (see Sections 2.3 and 2.4). Thus, one can synthesize RNA using any commercial DNA synthesizer with this minor modification to the standard 2-cyanoethyl phosphoramidite DNA cycle. However, in contrast with DNA oligonucleotides, the deprotection and purification of the chemically synthesized RNA requires care to avoid RNase contamination, which can cause rapid degradation of unprotected oligoribonucleotides. Wear disposable latex or plastic gloves and use only sterile glassware, plasticware, and buffers (see Chapter 1, Section 1.1).

For the preparation of biotinylated oligo(2′-O-alkylribonucleotides) for use in affinity procedures (Sections 3 and 4), single or multiple biotinylation may be performed during the solid-phase synthesis (4, 5). The incorporation of other non-radioactive reporter groups is also possible either during or after the oligomer assembly (6).

2.3 Chemical synthesis of oligoribonucleotides

Protocol 1 describes the synthesis of oligoribonucleotides using 2′-O-trialkylsilyl protection followed by their deprotection and purification by denaturing gel electrophoresis. Synthesis of the fully-protected oligoribonucleotide utilizes a similar methodology to that for an oligodeoxyribonucleotide, except that the coupling time is extended.

Since the 2′-O-trialkylsilyl protecting group is somewhat susceptible to cleavage by the 25% aqueous ammonia solution normally used to deacylate the N-protected nucleoside bases, a mixture of ethanolamine and ethanol (1:1 v/v) is used instead (7). This reagent not only removes the N-acyl protection of the nucleoside bases but also liberates the oligomer from the solid support and cleaves the 2-cyanoethyl protecting groups. In the next step, the 2′-O-trialkylsilyl protecting groups are removed by treatment with tetrabutylammonium fluoride in tetrahydrofuran. Note that subsequent removal of excess tetrabutylammonium fluoride is essential to prevent possible degradation of the product. This can be achieved on an anion-exchange cartridge (2), by gel filtration or, most easily, by butan-1-ol precipitation of the fully deprotected oligoribonucleotide.

Protocol 1. Synthesis, deprotection, and purification of oligoribonucleotides using 2′-O-trialkylsilyl protection

Equipment and reagents
- Automated solid-phase DNA synthesizer (for example Applied Biosystems) and ancillary equipment
- Speedvac centrifugal concentrator
- Denaturing 10% polyacrylamide/8 M urea gel and electrophoresis equipment (see *Protocol 11*)

Continued

Protocol 1. *Continued*

- 5'-*O*-dimethoxytrityl-2'-*O*-trialkylsilylribonucleoside-3'-*O*-(2-cyanoethyl-*N,N*-diisopropylphosphoramidites) under argon in septum-sealed vials and functionalized controlled pore glass supports (Milligen)
- 5 ml and 10 ml gas-tight syringes (Hamilton)
- Anhydrous acetonitrile (<30 p.p.m. water)
- Ethanolamine:ethanol (1:1, v/v)
- Butan-1-ol
- 1.1 M tetrabutylammonium fluoride in tetrahydrofuran (Aldrich; it is recommended that only small bottles of this reagent are purchased and are replaced at least once a month)
- TE buffer: 10 mM Tris–HCl, pH 7.5, 1 mM EDTA
- Glacial acetic acid: H_2O (1:19, v/v); sterilize by autoclaving
- 2.0 M Tris–acetate, pH 7.9, 1 mM EDTA; sterilize by autoclaving

Method

1. Prepare 0.1 M solutions of the ribonucleotide monomers by injecting the required volume of anhydrous acetonitrile through the septum cap using a dry gas-tight syringe.
2. According to the make of synthesizer being used, fill a synthesis cartridge or column with the appropriate functionalized controlled pore glass support to match the 3'-ribonucleoside of the oligoribonucleotide that you wish to synthesize. A scale of 0.5–1 µmol is sufficient for most purposes.
3. Following the instructions for the model of DNA synthesizer being used, assemble the desired sequence using the 0.1 M phosphoramidite solutions in anhydrous acetonitrile and a 2-cyanoethyl phosphoramidite cycle with the coupling time extended to 10 min.
4. Complete the oligoribonucleotide assembly then transfer the support bearing the fully-protected oligoribonucleotide to a screw-top vial. Add 1 ml of ethanolamine:ethanol, seal the vial, and keep it at 60 °C for 3 h.
5. Cool the vial to room temperature and let the insoluble support settle. Transfer the supernatant to a sterile microcentrifuge tube and lyophilize the protected oligonucleotide on a Speedvac.
6. Dissolve the oligonucleotide in 0.4 ml of 1.1 M tetrabutylammonium fluoride in tetrahydrofuran and mix well. Incubate the tube with the lid closed for 24 h at room temperature.
7. Transfer the solution as three equal portions to three sterile 1.5 ml microcentrifuge tubes. Add 1.2 ml butan-1-ol to each, close the tubes, and mix the contents by inversion.
8. Centrifuge the tubes for 5 min in a microcentrifuge to obtain the 5'-*O*-dimethoxytrityl protected oligonucleotide as a white or off-white pellet. Carefully remove and discard the supernatant.
9. Dissolve each pellet in 0.1 ml of sterile TE buffer.
10. Precipitate the 5'-*O*-dimethoxytrityl-protected oligoribonucleotide by the addition of 1 ml of butan-1-ol to each tube.

Continued

Protocol 1. *Continued*

11. Mix the components by inversion and recover the oligonucleotides as in step 7.
12. Dissolve each pellet in 50 µl of sterile acetic acid:water (1:19 v/v) and incubate the mixture at room temperature for 15 min to cleave the 5'-O-dimethoxytrityl protecting group.
13. Neutralize the contents of each tube by the addition of 0.1 ml of sterile 2 M Tris–acetate buffer, pH 7.9, 1 mM EDTA and then add 1.3 ml of butan-1-ol to each.
14. Cap the tubes, mix the contents by inversion, and centrifuge for 5 min in a microcentrifuge.
15. Carefully remove and discard the supernatant and dry the oligoribonucleotide pellet in a Speedvac.
16. Analyse an aliquot of the crude product by PAGE under denaturing conditions (see *Protocol 11*).

Using the procedure described in *Protocol 1*, the 'failure bands' observed by UV shadowing (i.e. less than full length products) are generally very weak, even for oligomers up to 30–40 nt long. This indicates that chemical synthesis is normally very efficient and the oligonucleotide usually does not require further purification. If necessary, however, the full length product can be further purified by preparative denaturing PAGE as described in Chapter 1, *Protocol 8*.

2.4 Chemical synthesis of oligo(2'-*O*-methylribonucleotides) and oligo(2'-*O*-allylribonucleotides)

The synthesis and applications of oligo(2'-*O*-methylribonucleotides) have recently been described in detail (4). The recently developed 2'-*O*-allyl analogues possess some additional advantages (8) and improved syntheses of the 2-cyanoethyl phosphoramidite monomers have been described (9). These oligomers are characterized by their resistance to both chemical and nuclease degradation.

Basically, the assembly of fully-protected support-bound oligo(2'-*O*-methylribonucleotides) and the 2'-*O*-allyl analogues (see *Figure 1*) is identical to that of oligoribonucleotides (described in *Protocol 1*). Although the authors have previously used 5-(4-nitrophenyl)-1-H-tetrazole as the condensing agent for synthesis of oligo(2'-*O*-alkylribonucleotides) (4, 8), the use of tetrazole gives equally good results. Suitably protected 2'-*O*-methylribonucleoside-3'-*O*-(2-cyanoethyl-*N,N*-diisopropylphosphoramidites) are already commercially available from Glen Research. 2'-*O*-Allylribonucleotide monomers are available from and 2'-*O*-allylinosine and 2'-*O*-allyl-2-aminoadenosine derivatives will be available from Boehringer Mannheim. A procedure for the chemical synthesis of oligo(2'-*O*-methylribonucleotides) and oligo(2'-*O*-allylribonucleotides) is described in *Protocol 2*. On completion of assembly, the oligonucleotides are deprotected initially by incubation overnight with 30% aqueous ammonia solution. This procedure cleaves the oligomer from the controlled pore glass support, removes the 2-cyanoethyl protecting groups and also the *N*-acyl protecting groups from the nucleoside bases. The crude 5'-*O*-dimethoxytrityl protected oligo(2'-*O*-alkylribonucleotide) is then conveniently purified by reversed phase HPLC on μ-Bondapak C_{18} (Waters), or similar material. Since depurination during the oligomer assembly is negligible, even for long A-rich sequences, the dimethoxytrityl-containing peak is generally homogeneous. The dimethoxytrityl protecting group is cleaved as described in *Protocol 2*. Further purification by PAGE is generally unnecessary. Alternatively, the final dimethoxytrityl protecting group may be cleaved before release of the oligomer from its solid-phase support in the ammonia step, but purification should then be achieved by PAGE (see *Protocol 11* or Chapter 1, *Protocol 8*) instead of HPLC.

Protocol 2. Synthesis of oligo(2'-*O*-methylribonucleotides) and oligo(2'-*O*-allyribonucleotides)

Equipment and reagents

- Automated solid-phase DNA synthesizer and ancillary equipment, Speedvac centrifugal concentrator, 5 ml and 10 ml gas-tight syringes and anhydrous acetonitrile as described in *Protocol 1*
- Gradient HPLC system and reversed-phase column (for example μ-Bondapak C_{18} column from Waters)
- 25 ml pear-shaped glass flasks
- 5'-*O*-Dimethoxytrityl-2'-*O*-methylribonucleoside-3'-*O*-(2-cyanoethyl-*N*,*N*-diisopropylphosphoramidites) under argon in septum-sealed vials and functionalized controlled pore glass supports (Glen Research)
- 5'-*O*-Dimethoxytrityl-2'-*O*-allylribonucleoside-3'-*O*-(2-cyanoethyl-*N*,*N*-diisopropylphosphoramidites) under argon in septum-sealed vials and functionalized controlled pore glass supports (Boehringer Mannheim)
- 0.1 M triethylammonium acetate buffer, pH 7.0 in distilled water
- 80% acetic acid (v/v)
- Diethyl ether
- Rotary evaporator

Method

1. Prepare 0.1 M solutions of the 2'-*O*-methyl- or 2'-*O*-allylribonucleoside monomers by injecting the required volume of anhydrous acetonitrile through the septum cap using a dry gas-tight syringe.
2. According to the model of synthesizer being used, fill a synthesis cartridge or column with the appropriate functionalized controlled pore glass support to match the 3'-nucleoside of the oligo(2'-*O*-alkylribonucleotide) that you wish to synthesize. A scale of 0.5–1 μmol is sufficient for most purposes.
3. Assemble the desired sequence using a 2-cyanoethyl phosphoramidite cycle with the coupling time extended to 6 min for 2'-*O*-methyl RNA and 10 min for 2'-*O*-allyl RNA using the 'trityl on auto' end procedure and using the standard condensing agent tetrazole.

Continued

Protocol 2. *Continued*

4. Seal the vial containing partially protected oligo(2'-O-alkylribonucleotide) in ammonia solution, and keep at 60 °C for 8–10 h (generally overnight) to remove the heterocyclic protection.
5. When cool, evaporate the solution to dryness at room temperature using a Speedvac.
6. Dissolve the residue in about 250 μl of 0.1 M triethylammonium acetate buffer and purify the 5'-O-dimethoxytrityl-oligo(2'-O-alkylribonucleotide) by reversed-phase HPLC on a μ-Bondapak C_{18} or equivalent column using a gradient of acetonitrile in 0.1 M aqueous triethylammonium acetate, pH 7.0. The oligonucleotide product will elute at about 35–40% acetonitrile.
7. Lyophilize the product fraction on a Speedvac.
8. Dissolve the residue in 500 μl of 80% acetic acid and keep at room temperature for 30 min to cleave the 5'-O-dimethoxytrityl group.
9. Add 500 μl of H_2O and transfer the whole solution to a 25 ml pear-shaped glass flask.
10. Add 5 ml of diethyl ether and mix well by shaking. Leave on the bench until the phases separate. The oligonucleotide remains in the lower (aqueous) phase. Discard the upper phase.
11. Repeat step 10 twice more.
12. Lyophilize the ether-extracted oligonucleotide solution (i.e. the aqueous phase from the final extraction) on a rotary evaporator.
13. Resuspend the oligonucleotide in 1 ml of H_2O. The oligonucleotide is now ready for use and its concentration can be determined by measuring A_{260} in a spectrophotometer ($A_{260} = 1.0$ is equivalent to an oligonucleotide concentration of 25 μg/ml).

3. Antisense affinity selection of RNAs or RNP complexes

3.1 Strategy

Specific RNAs or RNP complexes can be isolated directly from crude cellular extracts using an affinity selection approach with biotinylated antisense probes. For this application it is important to use antisense probes which form duplexes with the target RNA sequence characterized by high melting temperatures (T_m). It is also important that the resulting duplexes are not substrates for RNase H, which is commonly present in crude nuclear and cellular extracts.

The complex formed between the RNP particle and biotinylated antisense probe is purified on a solid support to which is bound either avidin, streptavidin, or anti-biotin antibodies. A wide range of such products are commercially available and suitable for use. Reproducibly reliable results have been obtained with streptavidin–agarose and so protocols based on this technology are described below (Section 3.3).

3.2 Probes for affinity selection

3.2.1 Advantages of substituted oligoribonucleotide probes

The use of either 2'-*O*-methyl or 2'-*O*-allyl substituted oligoribonucleotide probes is particularly recommended since they are nuclease resistant and form extremely stable, RNase H-resistant hybrids with complementary RNAs (8). The RNA:2'-*O*-alkylRNA heteroduplexes actually have significantly higher T_m values than the corresponding RNA:RNA hybrids of identical nucleotide sequence. These probes can be synthesized as described in Section 2. Lower backgrounds of non-specific probe binding are often obtained by using 2'-*O*-methylinosine in place of 2'-*O*-methylguanosine.

3.2.2 Biotinylation of probes

The most efficient method of incorporating biotin into the oligonucleotides is during synthesis by using a phosphoramidite substrate consisting of a modified base coupled to biotin via a flexible spacer arm (4, 5). An alternative, though less efficient, method is to incorporate an amino linker in the oligonucleotide and then to couple biotin to this linker after synthesis using a biotin ester such as sulphosuccinimidyl 6-(biotinamido)hexanoate [NHS-LC-biotin] (10). Suitable reagents for post-labelling probes can be purchased from Pierce. The best results have been obtained using probes incorporating two to four tandem biotin residues at either the 3' or 5' terminus of the oligonucleotide. However, since the position of biotin residues can, in certain cases, significantly affect the efficiency of affinity selection of targeted complexes, presumably due to steric hindrance from proteins in the RNP particle (11), it is better to incorporate two biotin residues at each end of the probe.

Protocol 3. Solid-phase synthesis of biotinylated probes

Equipment and reagents

- Equipment and reagents listed in Protocol 2
- Biotin-coupled phosphoramidite, e.g. 5'-*O*-dimethoxytrityl-N^4-methyl, N^4-[*N*-methyl, *N*-(*N*-{4-tert.-butylbenzoyl}biotinyl)-8-amino-3,6-dioxaoctyl] 2'-deoxycytidine-3'-*O*-(2-cyanoethyl *N,N*-diisopropylphosphoramidite) (MWG-Biotech) or biotin-dT (Glen Research)

Method

1. Follow step 1 of *Protocol 2* and in addition prepare a 0.1 M solution of the biotin-containing phosphoramidite in anhydrous acetonitrile. Place the biotin reagent in a spare monomer position on your synthesizer.

Continued

> **Protocol 3.** *Continued*
>
> 2. Use a 3′-terminal dT controlled pore glass support to start your synthesis, first couple on two biotin-containing monomers then synthesize your desired sequence as in step 3 of *Protocol 2*. Now add on two more biotin-containing monomers before performing the 'trityl on auto' end procedure. If the oligonucleotide being made needs to be efficiently 5′ phosphorylated (for example, for 5′ labelling with ^{32}P-phosphate) it is recommended that an additional 5′ nucleotide (such as dT) is added to the probe after the biotinylated residues. In this case add a dT monomer before performing the 'trityl on auto' end procedure.
> 3. Follow steps 4–10 of *Protocol 2* to obtain a tetrabiotinylated 2′-*O*-alkyl RNA probe. Note however that biotinylated oligonucleotides will elute at a slightly higher acetonitrile concentration than unbiotinylated oligonucleotides of otherwise identical sequence.

An alternative, but less convenient, procedure is postsynthesis biotinylation. However, this requires the incorporation of amino-modifier phosphoramidites instead of the biotin-containing monomer. Suitable reagents are the amino modifier-dT available from Glen Research or the amino modifier II from Cruachem. The oligonucleotide thus synthesized and purified containing primary amino groups is readily biotinylated using a suitable activated ester of biotin (10).

3.2.3 Nucleotide sequence and design of probes

There is no simple rule for knowing which region of an RNA will be the best target for an antisense oligonucleotide. In general, there is a good correlation between regions susceptible to nuclease cleavage or chemical modification and efficient probe binding (12–14). All RNase H sensitive sites are good candidates for targeting biotinylated probes. Note, however, that 2′-*O*-alkyl RNA probes can also bind stably to target sequences which are not, or are only poorly, cleaved by RNase H. This is probably the result of both the higher stability of RNA:2′-*O*-alkyl RNA hybrids as opposed to RNA:DNA hybrids and the inaccessibility of the bulky RNase H enzyme to certain sites of hybrid formation. Sequences known to be binding sites for proteins are not likely to be suitable targets for antisense probes. In contrast, RNA sequences predicted to form stems can sometimes be successfully targeted, since, as mentioned in Section 2.1, antisense probes may be able to displace intramolecular secondary structures due to the higher stability of the 2′-*O*-alkyl RNA:RNA hybrids compared to RNA:RNA duplexes (13).

Since the 2′-*O*-alkyl oligoribonucleotides form very stable hybrids with RNA, it is usually not essential to use long probes to obtain efficient affinity selection. Very good results can often be achieved with probes in the size range 11–16 nt, depending on the base composition of the target sequence. The stability of short hybrids can be additionally increased by including the modified base 2-aminoadenine in the oligonucleotide to base pair with uracil in the target sequence (14). Three hydrogen bonds are formed between 2-aminoadenine and uracil instead of the usual two hydrogen bonds formed by adenine. As discussed above, in the interest of minimizing background it can often be worth replacing 2'-O-alkylguanosine residues with 2'-O-alkylinosine. However, in cases where only a short region of the RNA target is available for interaction with the antisense probe, guanosine should be retained.

3.3 Biotin–streptavidin affinity selection of RNP complexes

3.3.1 Preblocking streptavidin–agarose beads

Affinity chromatography of RNP complexes (or RNAs) bound to biotinylated antisense oligonucleotides requires highly specific, biotin-dependent binding to the streptavidin–agarose beads. Therefore, to reduce non-specific binding, the beads should be preblocked shortly before use. This is described in *Protocol 4*.

Protocol 4. Preparation of preblocked streptavidin–agarose beads

Equipment and reagents

- Rotating wheel stirrer
- Streptavidin–agarose beads (normally supplied as a slurry in 10 mM sodium phosphate, pH 7.2, containing 0.15 M NaCl and 0.02% sodium azide); Sigma streptavidin–agarose is particularly recommended
- Preblock buffer (20 mM Hepes–KOH, pH 7.9, 0.3 M KCl, 0.01% Nonidet P-40, 50 µg/ml glycogen, 0.5 mg/ml BSA, 50 µg/ml yeast tRNA)
- Wash buffer (20 mM Hepes–KOH, pH 7.9, 50 mM KCl, 0.1% Nonidet P-40)

Method

1. Centrifuge the beads in a microcentrifuge at 4000 r.p.m. (1400 g) for 20 sec. Note that centrifuging the agarose beads too hard or too long can damage them and so should be avoided. This also applies to all subsequent washing steps.
2. Carefully remove and discard the supernatant and note the packed bead volume.
3. Gently resuspend the beads in an equal volume of preblock buffer.
4. Stir the beads for 15 min at 4 °C on the rotating wheel stirrer.
5. Collect the beads by centrifugation in a microcentrifuge at 4000 r.p.m. (1400 g) for 20 sec.
6. Resuspend the beads in an equal volume of wash buffer and then centrifuge them again as in step 5. Repeat this washing step twice more.
7. Remove the supernatant wash buffer from the beads after the final wash step. Keep the preblocked beads on ice until ready for use. They can be kept like this for at least several hours without loss of activity.

3.3.2 Affinity selection procedure

Protocol 5 describes the affinity selection of RNP complexes from a cell or nuclear extract using a biotinylated oligo(2'-O-alkylribonucleotide) targeted to the RNA component of the complex. The following points should be noted:

(a) The optimal concentration of antisense probe required to select a particular RNP complex depends upon the abundance of the RNP in the cell extract and the accessibility of the targeted complementary region of the RNA. Ideally the optimum concentration of the probe required should be determined in a pilot titration experiment. However, in practice approximately 0.2–0.5 pmol of biotinylated antisense probe per microgram of total protein in the extract is generally sufficient for efficient selection of even abundant RNP complexes. Note that if high concentrations of oligonucleotide are used, it may be necessary to increase the quantity of streptavidin–agarose beads in the selection. This can also be checked in the pilot titration experiment; if there is a decrease in selection efficiency when the probe concentration is increased, then it is very likely that the amount of streptavidin–agarose beads used is too low.

(b) The optimum time for incubation of the probe with the cell extract may also vary for different targeted RNPs. As a general guide, a 30–60 min incubation at 30 °C is usually sufficient.

(c) Incubation at lower temperatures (i.e. below 20 °C) may reduce selection efficiency.

(d) It is advisable to incubate the probe with the cell extract in the presence of ATP and creatine phosphate since this can help prevent protein aggregation and precipitation. The mechanism responsible for this preventative effect of ATP is not clear. The cell extract normally contains sufficient endogenous creatine kinase, so supplementation with purified enzyme is not required.

Protocol 5. Affinity selection of RNA complexes using biotinylated oligonucleotides

Equipment and reagents

- Rotating wheel stirrer
- 3 M KCl
- 15 mM ATP
- 50 mM creatine phosphate
- 20 mM $MgCl_2$
- WB300 (20 mM Hepes–KOH, pH 7.9, 0.3 M KCl, 0.1% Nonidet P-40)
- Preblocked streptavidin–agarose beads (see *Protocol 4*, prepared fresh and stored on ice)
- Biotinylated antisense oligonucleotide probe (see *Protocols 2* and *3*)
- Cell or nuclear extract (normally at 10 mg protein/ml), for preparation of extracts see Chapters 3 and 6, and Volume II, Chapter 3.

Continued

Protocol 5. *Continued*

Method

Note: If a pilot titration experiment is to be performed to determine the concentration of probe required, set up a series of 100 µl reactions according to step 1, but vary the amount of antisense oligonucleotide, typically over a range from 50–500 pmol. Make up the volume of each reaction to 100 µl with sterile H_2O. Proceed as in the remaining steps of the Protocol. In the subsequent experiment, use the concentration of probe found to give the maximum recovery of RNP complexes.

1. In a 1.5 ml microcentrifuge tube, mix:

    - cell or nuclear extract — volume containing 0.5 mg protein
    - biotinylated oligonucleotide probe — volume containing 200 pmol
    - 3 M KCl — 10 µl
    - 15 mM ATP — 10 µl
    - 50 mM creatine phosphate — 10 µl
    - 20 mM $MgCl_2$ — 10 µl
    - sterile H_2O — to 100 µl final volume

2. Incubate the mixture for 45–60 min at 30 °C.

3. Centrifuge in a microcentrifuge at 12000 r.p.m. (12700 g) for 3 min to remove any aggregated material which may have formed during the incubation. This step can significantly reduce the non-specific background.

4. Transfer the supernatant to a fresh microcentrifuge tube and discard the pellet.

5. Gently centrifuge down the preblocked streptavidin–agarose beads at 4000 r.p.m. (1400 g) for 20 sec and discard all of the supernatant.

6. Add an equal volume of preblocked streptavidin–agarose beads (i.e. packed bead volume) to the supernatant from step 4 and stir the suspension at 4 °C for 45–60 min using the rotating wheel stirrer. It may be necessary to add a larger quantity of streptavidin–agarose beads if the oligonucleotide probe is used at high concentration (see the text).

7. Recover the beads by centrifugation at 4000 r.p.m. (1400 g) for 20 sec in a microcentrifuge and remove the supernatant. The supernatant can either be discarded or retained for RNA analysis (see *Protocol 6*) to determine the efficiency of selection of the targeted RNP.

8. Gently resuspend the streptavidin–agarose beads in 2–3 vol. of WB300.

9. Stir the suspension for 5 min at 4 °C and recover the beads as in step 7.

10. Repeat steps 8 and 9 twice. Further washing generally does not improve the ratio of the selected RNP complex to the non-specific background.

11. Elute the affinity-selected RNP complex from the beads (see *Protocol 6*).

4: RNP complexes

3.3.3 Recovery of affinity-selected RNA

The method of elution of affinity-selected RNA from streptavidin–agarose beads will be dictated by the type of analysis which is to be carried out subsequently. *Protocol 6* describes recovery of RNA components from streptavidin–agarose using proteinase K/phenol. The analysis of these RNAs is described in Section 7, whereas the analysis of proteins from affinity-selected complexes is covered in Section 5.

Protocol 6. Recovery of affinity-selected RNA from streptavidin–agarose beads

Reagents

- Streptavidin–agarose with bound affinity-selected RNP complexes (or RNAs) prepared as in *Protocol 5*, step 10
- Proteinase K (20 mg/ml)
- Glycogen (0.2 mg/ml)
- PK buffer (0.1 M NaCl, 10 mM Tris–HCl, pH 7.6, 1 mM EDTA, 0.5% SDS)
- Phenol:chloroform (1:1 v/v)
- TE buffer (10 mM Tris–HCl, pH 7.5, 1 mM EDTA)

Method

1. Centrifuge the washed streptavidin–agarose beads (from *Protocol 5*, step 10) in a microcentrifuge for 20 sec at 4000 r.p.m. (1400 g). Remove the supernatant and resuspend the beads in 0.3 ml of PK buffer.
2. Add 35 µl of proteinase K and 20 µl of glycogen stock solutions.
3. Incubate the mixture at 65 °C for 45 min.
4. Pellet the beads by centrifuging the mixture in a microcentrifuge for 1 min at 4000 r.p.m. (1400 g).
5. Transfer the supernatant to fresh tube, taking care not to transfer any streptavidin–agarose beads, and extract it by vortexing it with an equal volume of phenol:chloroform.
6. Centrifuge the mixture for 3 min at 10 000 r.p.m. (8800 g). Collect the upper (aqueous) phase and mix with 3 vol. of absolute ethanol to precipitate the RNA. There is no need to add additional salt at this stage. Leave either overnight at −20 °C or 1 h in a dry ice/ethanol bath.
7. Recover the RNA precipitate by centrifuging for 10 min in a microcentrifuge at 12 000 r.p.m. (12 700 g).
8. Wash the precipitate twice with 0.5 ml of 70% ethanol, dry briefly, and resuspend the washed RNA in 25 µl of TE buffer for storage. Alternatively, redissolve the RNA directly in the appropriate gel loading buffer for immediate analysis as in Section 7.

An example of the affinity selection of an snRNA from a HeLa cell nuclear extract is shown in *Figure 2* (lane 3).

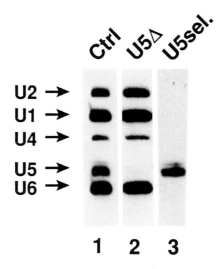

Figure 2. Antisense affinity depletion and affinity selection of U5 snRNP. A Northern blot is shown which has been probed for each of the five snRNA species (U1, U2, U4, U5, and U6) known to be present in mammalian spliceosomes. Lane 1 (Ctrl); RNA from a mock-depleted (control) HeLa cell nuclear extract (see *Protocol 7*). Lane 2 (U5 Δ), HeLa cell nuclear RNA after specific antisense affinity depletion of the U5 snRNP (see *Protocol 7*). Lane 3 (U5 sel.), affinity-selected U5 snRNA isolated using a U5-specific antisense probe (see *Protocol 5*). In each case the RNA was eluted from the streptavidin–agarose beads after selection (see *Protocol 6*). All the RNA samples were then analysed by denaturing PAGE (see *Protocol 11*) and electroblotted onto Hybond membrane (see *Protocol 13*). Probe preparation and Northern hybridization were carried out as described in *Protocols 14* and *15*.

4. Depletion of cell extracts of RNAs or RNP complexes using affinity selection with antisense oligonucleotide probes

In order to assess the functional role of individual RNAs or RNP complexes, it is useful to be able to study extracts in which a single RNA or RNP has been selectively removed. In many cases this can be done efficiently using biotinylated antisense probes to remove the targeted complex (15). The principle of the technique is analogous to that described in Section 3 for antisense affinity selection of RNPs or RNAs. An important difference, however, is that an extremely high depletion efficiency must be achieved. This contrasts with the affinity selection procedure where it is more important to achieve a high specificity with low non-specific background, than to obtain total recovery. On the other hand, in the depletion procedure a higher background of non-specific depletion can be tolerated to ensure that as much as possible of the targeted species is removed.

4: RNP complexes

Protocol 7 was originally developed for use with HeLa cell nuclear extracts (see Chapter 3). With minor modifications it can be adapted for use with extracts from other types of cell. Note that:

(a) It is important to carry out the affinity selection under high salt conditions and so the extract is dialysed against MD 0.6 buffer during the preparation procedure.

(b) As for affinity selection (see *Protocol 5*), the concentration of antisense oligonucleotide is best determined in a pilot experiment, but will often be in the range of 1–4 nmol per millilitre of nuclear extract.

(c) In order to optimize depletion efficiency, the authors recommend using antisense probes synthesized with guanosine and not inosine. It can also be helpful in certain cases to replace adenosine with 2-aminoadenosine, which results in more stable hybrids between the probe and the targeted RNA (14).

(d) For optimum depletion efficiency it may be necessary to repeat the affinity selection procedure by adding a fresh aliquot of preblocked streptavidin–agarose beads to the nuclear extract and incubating again for 45 min at 0 °C (see *Protocol 7*, step 8). Better depletion is obtained by performing two rounds of selection for 45 min each than by initially adding twice the volume of beads and incubating for 90 min. In general, it is important to use the minimum amount of streptavidin–agarose beads which still gives efficient depletion, since incubation with an excess of streptavidin–agarose beads can non-specifically inhibit enzymatic activity in the extract.

Protocol 7. Antisense affinity depletion of nuclear extracts

Equipment and reagents

- Rotating wheel stirrer
- 15 ml sterile plastic tubes
- 40 ml Dounce homogenizer (all glass type) with type B pestle
- Preblocked streptavidin–agarose beads (see *Protocol 4*)
- Biotinylated antisense probe (use 2′-O-alkyloligoribonucleotides; see *Protocols 2* and *3*)
- HeLa cell nuclei (for preparation see Chapter 3 and Volume II, Chapter 3)
- 0.1 M ATP
- 0.5 M creatine phosphate
- 5% Nonidet P-40
- 1 M DTT stock solution (store frozen in 1 ml aliquots)

- 0.1 M PMSF stock solution
- S buffer (20 mM Hepes–KOH, pH 7.9, 10% glycerol, 0.42 M KCl, 1.5 mM MgCl$_2$, 0.2 mM EDTA, 0.5 mM DTT, and 0.5 mM PMSF)
- MD 0.1 buffer (20 mM Hepes–KOH, pH 7.9 10% glycerol, 0.1 M KCl, 0.2 mM EDTA, 0.5 mM DTT, and 0.5 mM PMSF)
- MD 0.6 buffer (MD 0.1 buffer containing 0.6 M KCl in place of the 0.1 M KCl)
- D (storage) buffer (20 mM Hepes–KOH, pH 7.9, 20% glycerol, 0.1 M KCl, 0.2 mM EDTA, 0.5 mM DTT, and 0.5 mM PMSF)

Continued

Protocol 7. *Continued*

Note: Prepare the buffers in advance but without DTT and PMSF. Autoclave them and then allow them to cool overnight in a refrigerator to 4 °C and then add DTT and PMSF immediately before use.

Method

1. Resuspend the HeLa cell nuclei in S buffer, using 4–4.5 ml per 10^9 nuclei. Transfer to a 40 ml Dounce homogenizer and lyse the nuclei with 8–10 strokes of a B-type pestle.
2. Transfer the lysed nuclei to 15 ml sterile plastic tubes. Extract the lysed nuclei by rotating the tubes slowly on a rotary wheel stirrer at 4 °C for 30 min.
3. Dialyse the resulting nuclear extract three times at 4 °C. Use 40 vol. of MD 0.1 buffer each time and change the buffer at 90 min intervals.
4. Remove the precipitate that will have formed in the extract by centrifugation at 4 °C for 10 min at 20 000 g.
5. Dialyse the supernatant three times at 4 °C. Use 40 vol. of MD 0.6 buffer each time and change the buffer at 40 min intervals.
6. Incubate the dialysed nuclear extract with the biotinylated probe. To do this, divide the extract into aliquots of 0.9 ml. Add the antisense probe in a cocktail containing ATP, creatine phosphate, and Nonidet P-40. The total volume of this cocktail should be not more than 10% of the total volume of nuclear extract (i.e. 100 µl), resulting in a final concentration of 1.5 mM ATP, 5 mM creatine phosphate and 0.05% Nonidet P-40, and a final volume of 1.0 ml. As explained in the text, the concentration of antisense oligonucleotide is best determined in a pilot experiment but will usually be in the range of 1–4 nmol of probe per millilitre of nuclear extract. Therefore, for each 0.9 ml of extract to be depleted, mix:

    - 0.1 M ATP   15 µl
    - 0.5 M creatine phosphate   10 µl
    - 5% Nonidet P-40   10 µl
    - biotinylated antisense probe   volume equivalent to 1–4 nmol
    - H_2O to 100 µl final volume

7. Incubate the probe/extract mixture at 30 °C for 60–90 min. The optimal time for incubation is best determined empirically in pilot experiments. Meanwhile, prepare the preblocked streptavidin–agarose beads. Each 1 ml aliquot of incubated extract (step 6) will require 0.5 ml of packed beads in a 2 ml microcentrifuge tube. Note however that if high probe concentrations are used it may be necessary to increase correspondingly the quantity of streptavidin–agarose beads used at this step. To avoid dilution of the extract when the beads are added, centrifuge the preblocked beads for 2 min at 4000 r.p.m. (1400 g) in a microcentrifuge and completely remove the supernatant before adding the beads to the extract. Take care not to centrifuge the beads too hard or they may be damaged.

Continued

4: RNP complexes

Protocol 7. Continued

8. Add the 1 ml aliquots of incubated extract (from step 6) to the tubes containing the 0.5 ml of packed beads. Slowly stir the samples at 0 °C for 45 min. It is very important to keep the temperature low; place each tube containing the nuclear extract and streptavidin–agarose beads within a larger tube packed with ice.
9. Remove the beads (after 45 min) by centrifuging at 4000 r.p.m. (1400 g) for 2 min in a microcentrifuge. Note it may be necessary to repeat steps 8 and 9 using fresh streptavidin–agarose beads, as explained in the text.
10. Pool the depleted extracts. Then dialyse the pooled extract into the required storage buffer (e.g. D buffer is suitable for most extracts).
11. After dialysis snap-freeze the depleted extract in liquid nitrogen in aliquots of 50–100 µl. Store the aliquots in liquid nitrogen or below −90 °C.

An example of a HeLa cell nuclear extract specifically depleted of U5 snRNP by this method is shown in *Figure 2* (lane 2). It is possible to obtain depletion efficiencies of over 99% using the method described in *Protocol 7*.

5. Analysis of proteins in RNP complexes purified by affinity selection with oligonucleotide probes

5.1 Introduction

Several groups have recently reported the isolation and characterization of proteins present in RNP particles using antisense probes made of 2′-O-alkyl RNA (16–19). This approach has been successfully applied to both mammalian and trypanosome nuclear extracts. The antisense oligonucleotide technique can be particularly useful in cases where no antibodies are available that recognize an RNP complex, or when it proves difficult to isolate the complex in large quantities by conventional chromatographic methods. Since the antisense affinity selection purifies RNP complexes by a completely different route to immunoaffinity chromatography, it also offers an alternative approach to confirming the protein composition of individual RNP complexes and may detect proteins whose association is not sufficiently stable for their isolation by immunological techniques. It should be emphasized, however, that the optimal situation is when RNP proteins can be characterized using *both* antisense oligonucleotide affinity chromatography and immunoaffinity chromatography in parallel. Immunoaffinity methods are described in Chapter 5.

5.2 Antisense affinity selection of RNP complexes from precleared cell extracts

This Section describes an antisense method, based on the procedure described by Blencowe (20), which allows the rapid identification of proteins present in individual RNP complexes isolated specifically by antisense affinity selection. However, unlike the procedure described in *Protocol 4*, whose purpose is the isolation of RNP complexes for subsequent analysis of their RNAs (or the direct isolation of RNAs themselves from extracts), the prime aim of the procedure described in *Protocol 9* is to be able to analyse RNP proteins directly from crude extracts, using a single antisense affinity selection step, while avoiding any additional purification. In order to achieve this, it is essential to have as low a background as possible. To minimize background, extensively 'preclear' the extract of proteins and RNPs which can stick to the beads non-specifically. *Protocol 8* describes this procedure. In addition, the signal:background ratio can be further improved by using antisense probes incorporating inosine in place of guanosine. It is convenient to prepare precleared extract in relatively large quantities and store it in 1 ml aliquots. Using the same batch of precleared extract facilitates a direct comparison of the results obtained with different probes.

Protocol 8. Preparation of precleared cell extracts

Equipment and reagents

- Rotating wheel stirrer and streptavidin–agarose beads as described in *Protocol 4*
- Cell or nuclear extract (typically 10 mg/ml total protein concentration), either prepared fresh or a frozen extract prepared previously (for preparation see Chapters 3 and 6); if using prefrozen extract, thaw it slowly on ice just prior to use, keep the extract at 0–4 °C
- 1.0 M DTT stock solution (store frozen in 1 ml aliquots)
- Buffer A (20 mM Hepes-KOH, pH 7.9, 0.1 M NaCl, 0.05% Nonidet P-40, and 0.5 mM DTT). Add the DTT from the 1 M stock solution just before use
- Buffer B (same as Buffer A but containing 0.25 M NaCl, instead of 0.1 M NaCl)

Method

Work at 0–4 °C during the following steps.

1. Wash the streptavidin–agarose beads twice with an equal volume each time of buffer B as described in *Protocol 4*, steps 1 and 2. After washing carefully remove and discard excess buffer supernatant. NB *Do not preblock these beads because the aim is to have them absorb as much as possible of the non-specific 'sticky' protein in the extract.*

2. Centrifuge 1 ml aliquots of the cell or nuclear extract in a microcentrifuge for 5 min at 12 000 r.p.m. (12 700 g) and carefully remove the supernatants into fresh 2 ml microcentrifuge tubes. Discard the pellets.

3. Dilute each supernatant with an equal vol. of buffer A. Add 0.25 vol. of the washed beads (packed bead volume) from step 1 to each tube.

4. Stir the samples slowly on the rotary stirrer for 1 h at 4 °C.

5. Centrifuge for 20 sec at 4000 r.p.m. (1400 g) in a microcentrifuge.

6. Remove the supernatant and add it to a fresh aliquot of washed beads in another tube. Repeat step 4.

Continued

Protocol 8. Continued

7. Repeat steps 5 and 6 at least 5 times.
8. After the final incubation, centrifuge the extract for 3 min at 4000 r.p.m. (1400 g) in a microcentrifuge.
9. Remove the supernatant into a fresh tube, but without beads, and centrifuge again as in step 8 to ensure complete removal of the beads.
10. Use the extract directly for protein analysis (see *Protocol 9*) or snap-freeze it in aliquots in liquid nitrogen and store the aliquots below −80 °C for future use.

Protocol 9. Purification of RNP proteins from precleared cell extracts

Equipment and reagents

- Microdialysis cups (1 ml size, molecular weight cut off 12 000; Sartorius)
- Speedvac centrifugal concentrator
- Rotating wheel stirrer, precleared cell or nuclear extract and buffer B (see *Protocol 8*)
- Streptavidin–agarose beads (prewashed in buffer B as in *Protocol 8*, step 1)
- Biotinylated antisense probe (see *Protocols 2* and *3*)
- 0.5 M creatine phosphate, 0.1 M ATP, 1 M DTT (see *Protocol 7*)
- 0.2 M $MgCl_2$
- Wash buffer (20 mM Tris–HCl, pH 7.6, 0.01% Nonidet P-40 and 0.25 M NaCl)
- 5.0 M NaCl
- 8.0 M urea

Method

1. Use the required number of 1 ml samples of precleared extract from *Protocol 8*. Adjust the NaCl concentration of each sample of extract to 0.25 M (final concentration) by adding 30 µl of 5 M NaCl.
2. Make up probe cocktail (sufficient for 0.9 ml extract) by mixing:
   - 0.1 M ATP — 15 µl
   - 0.5 M creatine phosphate — 10 µl
   - 0.2 M $MgCl_2$ — 10 µl
   - biotinylated antisense probe — volume containing 1–5 nmol
   - H_2O — to 100 µl final volume
3. Mix 100 µl of the cocktail with each 0.9 ml of the extract. Mix gently and incubate for 60–90 min at 30 °C.
4. After incubation, centrifuge the extract in a microcentrifuge at 12 000 r.p.m. (12 700 g) for 5 min to remove any precipitate which may have formed.
5. Transfer the supernatant to a fresh tube containing 0.2 ml packed volume of prewashed streptavidin–agarose beads.
6. Slowly stir the suspension on the rotatory stirrer for 90 min at 4 °C.

Continued

Protocol 9. Continued

7. Centrifuge down the beads in a microcentrifuge for 20 sec at 4000 r.p.m. (1400 g).
8. Remove the supernatant and resuspend the beads in 0.9 ml of wash buffer. Stir on the rotatory stirrer for 5 min at 4 °C.
9. Repeat steps 7 and 8 four more times.
10. After the final wash, remove 0.2 ml of the bead suspension and isolate the affinity-selected RNA as described in *Protocol 6*. This will be used to evaluate the efficiency and specificity of the selection procedure.
11. Recover the RNP proteins from the remainder of the bead suspension. To do this centrifuge down the beads as in step 7. Resuspend beads in 0.8 ml of 8 M urea. Incubate the suspension for 30 min at room temperature.
12. Centrifuge for 1 min at 10 000 r.p.m. (8800 g) in a microcentrifuge, and carefully remove the supernatant into a fresh tube. Repeat the centrifugation to ensure that all the streptavidin–agarose beads have been removed.
13. Remove the urea from the resulting supernatant by dialysis in a 1 ml microdialysis cup for 3–4 h against 3 litres of distilled H_2O.
14. After dialysis, add 30 μl of 10% glycerol to each sample and concentrate them overnight under vacuum in the Speedvac.

RNP proteins isolated by *Protocol 9* may be analysed by SDS–PAGE. Protein bands are best detected by silver staining. Protocols for these procedures are provided in Chapter 5 (*Protocols 7* and *8*) and in another volume in this series (21).

6. Targeted RNase H cleavage of RNA

A powerful method for studying the functional requirement for individual RNAs or RNPs in biochemical reactions is to use targeted RNase H cleavage. RNase H is an ubiquitous nuclease activity which cells use to remove, from genomic DNA, the short stretches of RNA that act as primers for DNA polymerase. Since RNase H cleaves only the RNA strand in a DNA:RNA heteroduplex, specific cleavage of an RNA species can often be achieved *in vitro* by incubating with RNase H in the presence of a complementary DNA oligonucleotide. So long as the complementary DNA oligonucleotide hybridizes specifically to only the targeted RNA, and does not cross-hybridize with other RNAs, this method can be used to cleave RNAs selectively even in crude extracts (22–26). It can also be applied in certain assays *in vivo* and has, for example, been used particularly effectively to inactivate snRNPs in *Xenopus* oocytes (27, 28).

4: RNP complexes

Protocol 10 describes a method suitable for studying RNA function using oligodeoxyribonucleotide-directed RNase H cleavage. The protocol was originally developed for use with HeLa cell nuclear extracts. With minor modifications it can be used with extracts from other types of cell. The following points should be noted:

(a) RNase H will cleave RNA only in RNA:DNA heteroduplexes of at least 4 bp or longer.

(b) Single- or double-stranded RNA or DNA is not cleaved. RNase H will also not cleave RNA that is hybridized to 2'-*O*-methyl or 2'-*O*-allyl oligoribonucleotides.

(c) Efficient RNase H cleavage requires an accessible, single-stranded region in the target RNA that can hybridize with a DNA oligonucleotide. Regions of RNAs that are bound by proteins or that form stable secondary structures are unlikely to be good targets for RNase H cleavage.

(d) It can also be helpful for efficient cleavage to use DNA oligonucleotides that are slightly longer (typically 16–20 nt) than required for affinity selection by antisense 2'-*O*-alkyl substituted oligoribonucleotides. This is a consequence of the lower thermal stability of DNA:RNA hybrids as compared with RNA:2'-*O*-alkyl RNA hybrids.

(e) Include ATP and creatine phosphate in the reactions (at final concentrations of 1.5 mM and 5 mM, respectively) since they increase the cleavage efficiency for some RNP complexes.

(f) The concentration of DNA oligonucleotide required should be determined by titration in a pilot experiment. If the concentration is too low, only partial cleavage will be obtained. Conversely, the use of very high concentrations can cause promiscuous cleavage of RNAs in the extract through cross-hybridization. Typically, oligonucleotide concentrations in the range of 10–50 μM (final concentrations) should be tested.

(g) Many extracts, particularly nuclear extracts, contain significant amounts of endogenous RNase H. It is, therefore, often possible to observe cleavage in the presence of DNA oligonucleotides without adding any exogenous RNase H to the reaction. However, with many extracts it is advisable to supplement the RNase H activity by adding additional enzyme.

(h) If the RNase H digestion is to be performed on purified RNA or RNP complexes, and not in an extract, it is of course essential to include exogenous RNase H activity. In this case the authors recommend that carrier RNA be included in the reactions to compete for any traces of general RNase activity in the RNase H preparation which would otherwise degrade the RNA under study. For this, yeast tRNA can be used at a final concentration of 20 μg/ml.

Protocol 10. Oligonucleotide-directed RNase H cleavage of RNA

Reagents

- Oligodeoxyribonucleotides (synthesized on an automated DNA synthesizer)
- RNase H (from *E. coli*; Boehringer Mannheim)
- Cell or nuclear extract, 50 mM creatine phosphate, 20 mM $MgCl_2$, 15 mM ATP as *Protocol 5*
- DNase (RNase-free DNase from Boehringer Mannheim is recommended)
- Proteinase K (20 mg/ml), PK buffer, TE buffer, and phenol:chloroform as *Protocol 6*

Method

1. Set up the following reaction in a microcentrifuge tube:

    - 15 mM ATP — 10 µl
    - 50 mM creatine phosphate — 10 µl
    - 20 mM $MgCl_2$ — 10 µl
    - cell or nuclear extract — volume containing 0.5 mg total protein
    - oligodeoxyribonucleotide — volume containing 2.5 nmol
    - H_2O — to 100 µl final volume

2. Mix gently and add 5–10 units of *E. coli* RNase H. Incubate the extract at 30 °C for 60 min.

3. If the extract is to be used for further functional experiments (for example, splicing assays), add 10 units of DNase (RNase free) to the reaction to remove the DNA oligonucleotide and prevent any additional cleavage of other RNAs taking place. Incubate for a further 15 min at 30 °C. Remove an aliquot to check cleavage efficiency and process it as described in steps 4–7. The remainder of the cleaved extract should be used immediately for the functional assay under study.

4. Analyse the efficiency of RNase H cleavage by centrifuging the sample in a microcentrifuge for 3 min at 12 000 r.p.m. (12 700 g) and discarding any pellet formed.

5. Transfer the supernatant to a fresh tube and add 0.2 ml of PK buffer and 0.1 ml of proteinase K solution. Incubate the mixture at 65 °C for 45 min.

6. Recover the RNA by deproteinization with phenol:chloroform followed by ethanol precipitation as described in *Protocol 6*, steps 5–7.

7. Wash the RNA with 70% ethanol and dissolve the RNA in either TE buffer (for storage) or directly in the appropriate gel loading buffer for immediate analysis by PAGE and Northern hybridization (see Section 7).

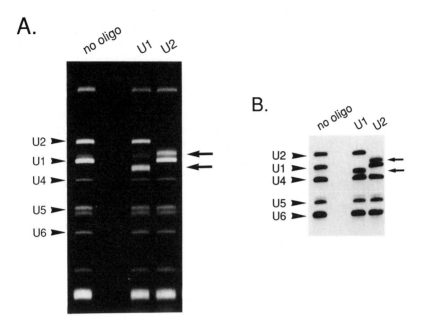

Figure 3. Oligonucleotide-directed RNase H cleavage of snRNAs in a HeLa cell nuclear extract. The effects of RNase H cleavage (see *Protocol 10*) are shown in extracts either in the absence of any oligodeoxyribonucleotide, or in the presence of oligodeoxyribonucleotides targeted to either the U1 or U2 snRNAs. The samples were then analysed by denaturing PAGE as described in *Protocol 11*. The RNAs were detected both by ethidium bromide staining (see *Protocol 11*; panel A) and by electroblotting followed by Northern hybridization (see *Protocols 13* and *15*; panel B). RNase H cleavage is demonstrated by the replacement of the targeted snRNA by a smaller product (lanes U1, U2). Cleavage is specific for the targeted snRNA since only the targeted snRNAs (U1 and U2) are affected by the RNase H and the enzyme has no affect when the oligodeoxyribonucleotides are absent (no oligo).

An example of DNA oligonucleotide-targeted RNase H cleavage of snRNA in a HeLa cell nuclear extract is shown in *Figure 3*.

7. Analysis of RNAs and RNPs by gel electrophoresis

7.1 Introduction

For most biochemical studies of RNA processing it is essential to analyse RNAs and RNP complexes by gel electrophoresis. Many alternative gel systems have been described for this purpose. In this section, protocols are given for both denaturing and native gels; the separation of affinity-selected RNAs uses denaturing polyacrylamide gels (Section 7.2) and the rapid separation of RNA complexes from crude nuclear extracts uses native polyacrylamide–agarose gels (Section 7.3). Protocols are also provided for the transfer of RNAs from both

7.2 RNA separation by denaturing polyacrylamide gel electrophoresis

The following gel system (see *Protocol 11*) can be used for analysing RNAs from affinity selection (see *Protocol 6*) or depletion (see *Protocol 7*), or for monitoring RNase H cleavage (see *Protocol 10*). This gel system is well suited to the analysis of nuclear RNAs, such as snRNAs, in the size range of 50–300 nt. For much smaller or much larger RNA species, the polyacrylamide concentration in the gel and time of electrophoresis should be adjusted appropriately. For quantitative Northern hybridization, also load different amounts of an RNA standard into separate lanes. After electrophoresis it is convenient to stain the gel with ethidium bromide to obtain a photographic record of the separation prior to electroblotting (Section 7.4) and Northern hybridization (Section 7.5).

Protocol 11. Separation of RNAs by denaturing PAGE

Equipment and reagents

- Electrophoresis equipment with 20 cm × 20 cm glass plates and 1.5 mm spacers and comb
- UV transilluminator
- Urea
- 40% (w/v) acrylamide:bisacrylamide (30:1) stock solution
- Ethidium bromide (10 mg/ml) (**NB. Carcinogen** — wear gloves when handling and store clearly labelled as toxic compound. Treat disposal of solutions containing ethidium bromide as chemical waste)
- RNA sample for analysis (for example, prepared as in *Protocols 6, 7* or *10*)
- 20 × TBE buffer (216 g Tris base, 110 g boric acid, 18.6 g EDTA, H_2O to 1 litre)
- 25% ammonium persulphate (freshly made)
- TEMED (BioRad)
- Formamide–dyes loading buffer; make this by mixing 2 ml of 0.5 M EDTA, 0.1 g bromophenol blue, 0.1 g xylene cyanol and deionized formamide[a] to 100 ml

Method

1. Assemble the gel plates and spacers using adhesive tape to seal the edges.
2. Make up 60 ml of 10% acrylamide, 8 M urea, TBE buffer as follows:
   - 40% acrylamide:bisacrylamide stock solution   15 ml
   - 20 × TBE buffer   3 ml
   - urea   27.6 g
   - H_2O to 60 ml final volume

Continued

Protocol 11. *Continued*

3. Add 0.1 ml of 25% ammonium persulphate and 0.1 ml of TEMED to the acrylamide gel mixture. Mix and pour it immediately between the gel plates. Insert the comb and clamp the gel sandwich in a vertical position during the polymerization process.
4. When the gel has polymerized, carefully remove the comb and assemble the gel sandwich on the electrophoresis apparatus using 1 × TBE buffer for both upper and lower reservoirs. Use a metal plate clamped to the front of the gel plates during electrophoresis to dissipate heat generated and so minimize distortion effects caused by uneven heat transfer. Carefully flush out the wells with TBE buffer immediately prior to loading samples.
5. If the RNA sample is in the form of a dried precipitate, dissolve it directly in formamide–dyes loading buffer. For RNA samples already in solution, mix them with 3 vol. of the formamide–dyes loading buffer. Heat the samples for 15 min at 65 °C and load them directly into the wells of the gel.
6. Run the gel at 500 V (constant voltage) until the bromophenol blue (i.e. the faster moving dye) reaches the bottom of the glass plates. The voltage may have to be altered for gels of differing thickness or acrylamide concentration.
7. Disassemble the apparatus and transfer the gel to a plastic or glass dish containing 500 ml of 1 × TBE buffer with 50 μl of ethidium bromide (10 mg/ml).
8. Gently shake the immersed gel at room temperature for 30 min.
9. Rinse the gel briefly in distilled H_2O and photograph the stained gel by UV transillumination to identify the RNA bands.

[a]Deionize the formamide as described in *Protocol 15*.

7.3 Separation of RNP complexes by electrophoresis in native polyacrylamide–agarose gels

Large RNP complexes, with the protein and RNA components still associated, can be separated by gel electrophoresis under native conditions. Many alternative gel systems have been described for this purpose and the choice of which system is optimal for a particular purpose depends on the properties of the complex, i.e. variables such as size and stability. *Protocol 12* describes a simple native gel system which will separate a range of RNP complexes formed in nuclear extracts, such as snRNPs and spliceosomes. The use of a composite polyacrylamide/agarose gel system is recommended, rather than a low percentage acrylamide gel, since the composite gels are mechanically stronger (less likely to break during handling) and, importantly, much easier to use for blotting analyses (Section 7.4).

Protocol 12. Separation of RNPs by native gel electrophoresis

Equipment and reagents

- Electrophoresis equipment, 25% (w/v) ammonium persulphate and TEMED as *Protocol 11*
- 1.0 M Tris–glycine buffer (121.1 g Tris base, 75 g glycine, H_2O to 1 litre)
- Electrophoresis buffer (75 mM Tris–glycine buffer made by dilution of the stock 1.0 M Tris–glycine buffer)
- 40% (w/v) acrylamide:bisacrylamide (60:1 w/w) stock solution
- Agarose (electrophoresis grade)
- Nuclear extract or other sample containing RNP complexes (for preparation of nuclear extracts see Chapter 3 and Volume II, Chapters 3 and 4)

Method

1. Assemble the gel plates and spacers using tape to seal the edges.

2. In a small flask, mix together 5.25 ml of the 40% acrylamide:bisacrylamide stock solution and 24.75 ml distilled H_2O.

3. In a separate flask, mix 0.8 g agarose, 15 ml of 1 M Tris–glycine buffer and 85 ml distilled H_2O. Heat the suspension on a hot plate until the agarose has dissolved.

4. Let the solution cool briefly and then add 30 ml to the flask with the acrylamide. (**NB** It is important to avoid adding very hot agarose to the acrylamide since this can impair the quality of the gel.) Immediately mix the two solutions together by swirling the flask.

5. Add 0.1 ml of 25% ammonium persulphate and 0.1 ml of TEMED, mix and immediately pour the solution between the gel plates. Insert the well-forming comb, clamp it and leave the gel to polymerize.

6. When the gel has polymerized, carefully remove the comb and assemble the gel sandwich on the electrophoresis apparatus.

7. Rinse the wells carefully with electrophoresis buffer immediately prior to loading samples. Samples containing nuclear extract are usually sufficiently dense that they can be loaded straight into the wells of the gel without requiring the addition of any loading buffer[a]. If, however, the protein concentration is very low, as, for example, with highly purified complexes, mix the sample with 0.25 vol. of 20% glycerol before loading.

8. Load the samples into the well of the gel and electrophorese at constant voltage (350 V). For the analysis of individual snRNP particles, run the gel for 90–120 min. For larger complexes, such as spliceosomes, extend the run to 5 h or longer. If extremely long runs are performed, it may be necessary to replenish the running buffer or to recirculate it.

Continued

Protocol 12. Continued

9. At the end of the electrophoresis, separate the gel plates, and visualize the resolved RNP complexes, either directly by autoradiography if labelled complexes were separated, or by transfer to a membrane for subsequent hybridization analysis. For transfer of complexes from the native gel to a membrane follow the electroblotting procedure described in *Protocol 13*.

[a] It can be helpful for monitoring the progress of the electrophoresis to load marker dyes on free lanes at the ends of the gel. Alternatively, dyes can also be added directly to the samples. Dye mix for native gels is prepared by mixing 20 ml glycerol, 80 ml H_2O, 0.1 g bromophenol blue, and 0.1 g xylene cyanol.

7.4 Transfer of RNAs from gels by electroblotting

Protocol 13 describes the transfer of RNAs from both denaturing polyacrylamide gels and native polyacrylamide–agarose gels.

Protocol 13. Electroblotting of RNAs from denaturing polyacrylamide gels or from native polyacrylamide–agarose gels

Equipment and reagents

- Electroblotting apparatus (e.g. from BioRad)
- UV crosslinking apparatus (the Stratalinker 2400 from Stratagene is particularly recommended)
- Hybond membrane (Amersham)
- 3 MM paper (Whatman)
- Transfer buffer (6.0 mM trisodium citrate, 8 mM Na_2HPO_4)
- Either native polyacrylamide–agarose gel with resolved RNP complexes and running buffer as in *Protocol 12* or denaturing polyacrylamide gel with resolved RNAs as in *Protocol 11*
- Electrophoresis buffer (see *Protocol 12*) for native gels only
- 50% urea, 25 mM Tris–glycine, 0.5 mM EDTA buffer (native gels only)

NB Wear disposable latex or plastic gloves when handling gels or Hybond membranes.

Method

A. *Gel preparation*

i. *Native gels*

1. Rinse the gel briefly in electrophoresis buffer.
2. Soak the gel, with gentle shaking, for 30 min at room temperature in 50% urea, 25 mM Tris–glycine, 0.5 mM EDTA.
3. Rinse the gel briefly three times in transfer buffer.

ii. *Denaturing gels*

1. Briefly rinse the gel in transfer buffer.

Continued

Protocol 13. *Continued*

B. *Electroblotting (all gels)*

1. Cut a piece of Hybond membrane to fit the area of the gel and three pieces of 3 MM paper of similar size. Wet the foam sponges from the electroblotting apparatus, the Hybond membrane, and the 3 MM paper thoroughly in transfer buffer.

2. Assemble the transfer sandwich of the electroblotting apparatus by placing the Hybond membrane directly on top of the gel surface. Take great care to exclude bubbles between the gel and the membrane; remove any that form using gloved fingers. Form the sandwich with (in order) foam sponge, 3 MM paper, Hybond membrane, gel, two layers of 3 MM paper, and finally foam sponge. To help prevent small bubbles forming between the layers of the sandwich, immerse the Hybond/gel/3 MM paper combination in transfer buffer before setting up the rest of the sandwich.

3. Place the assembled sandwich in the chamber of the electroblotting apparatus, orientating the sandwich such that the membrane side of the gel is facing the positive terminal. Fill the chamber completely with transfer buffer.

4. Follow (a) or (b)

   (a) *For denaturing gels*
   Transfer at 250 mA (constant current) for 60 min, followed by 350 mA (constant current) for a further 60 min.

   (b) *For native gels*
   Transfer at 250 mA (constant current) overnight at 4 °C.

5. Disassemble the sandwich after transfer, and place the wet Hybond membrane (gel side upwards) on a clean sheet of 3 MM paper. UV crosslink the RNAs to the membrane using the Stratalinker 2400. Do not let the membrane dry out before or during the crosslinking; keep it moist with transfer buffer.

6. Use the membrane at this stage for Northern hybridization (see *Protocol 15*) or store it for future analysis. For storage, place the membrane between two sheets of 3 MM paper and keep it in a plastic bag or wrapped in foil at room temperature.

7.5 Analysis of RNAs by Northern hybridization with riboprobes

Northern hybridization (see *Protocol 15*) is a convenient way to visualize specific RNAs separated by electrophoresis and electroblotted on to Hybond membrane as described in *Protocol 13*. Many different types of hybridization probe can be used. The authors have obtained good results using uniformly-labelled riboprobes. The preparation of ^{32}P-labelled riboprobes is described in *Protocol 14*, using a method whereby cloned DNA sequences are copied from linear plasmid templates by transcription with phage RNA polymerase. This requires the use of specialized vectors which contain phage promoters adjacent to the cloning sites, as described in Chapter 1 (Section 2.2). In Northern hybridization, the exact amount of riboprobe used may have to be varied for detecting different RNAs. The stringency of washing may also be varied according to the choice of probe and the RNAs to be detected; *Protocol 15* describes high stringency washing suitable for most applications.

In certain situations, multiple independent probes can be hybridized to a membrane simultaneously. The authors have had good results using as many as five separate riboprobes simultaneously. However, this should not be done indiscriminately with probes that have not previously been characterized. It is extremely important to establish that a probe gives the same signal when hybridized separately as when in combination with other probes. Care should always be taken to ensure that the probe is in excess if quantitative hybridization data are required. This can be checked by comparing the autoradiographic signals from the control RNA samples, loaded in parallel at different concentrations (see Section 7.2). The signals should increase in proportion to the amount of RNA loaded.

Protocol 14. Preparation of ^{32}P-labelled riboprobes

Reagents

- Linearized template DNA (e.g. Bluescript vector with the probe DNA sequence inserted into the polylinker)
- Phage T3 or T7 RNA polymerase[a]
- 10 × transcription buffer (0.4 M Tris–acetate, pH 7.9, 80 mM magnesium acetate, 0.5 M sodium acetate, pH 7.0, 20 mM spermidine)
- 0.1 M DTT (stored in aliquots at −20 °C)
- RNase inhibitor (e.g. RNasin; Promega)
- rNTP mixture (containing rATP, rCTP, and rGTP each at 2 mM, and 0.1 mM rUTP, final concentrations)[b]
- [α-^{32}P]UTP (Amersham; 20 mCi/ml, 800 Ci/mmol)
- DNase (RNase-free DNase from Boehringer Mannheim is recommended)
- TE buffer (10 mM Tris–HCl, pH 7.5, 1 mM EDTA)
- Sephadex G50 (equilibrated in TE buffer)
- Phenol:chloroform (1:1, v/v)

Method

1. Set up the following reaction in a 1.5 ml microcentrifuge tube:

    - 10 × transcription buffer   10 µl
    - rNTP mix   10 µl
    - 0.1 M DTT   10 µl
    - [α-^{32}P]UTP   6 µl
    - RNase inhibitor   volume containing 100 units
    - linear template DNA   volume containing 0.5 µg
    - H$_2$O   to 100 µl final volume

Continued

Protocol 14. Continued

2. Mix the reactants together and centrifuge the tube for 10 sec in a microcentrifuge at 10 000 r.p.m. (8800 g) to remove air bubbles and ensure the contents are at the bottom of the tube.
3. Start the reaction by adding 1 µl of phage RNA polymerase (50 units/µl) and incubate for 30 min at 37 °C.
4. Add a further 1 µl of phage RNA polymerase and continue the incubation for another 30 min at 37 °C.
5. Add 2.5 µl of DNase (10 units/µl) and incubate for 30 min at 37 °C.
6. Add an equal volume of phenol:chloroform (1:1 v/v), vortex for 30 sec and spin for 3 min in a microcentrifuge at 10 000 r.p.m. (8800 g).
7. Remove the upper (aqueous) phase and load it on to a 1 ml column of Sephadex G50, eluting the column in TE buffer. The labelled probe will elute from the column with the void volume and in this way will be separated from the unincorporated rNTPs.

NB. The probe is now ready for use in hybridization experiments. However, to check that a clean preparation of full-length probe has been generated, run a small aliquot on a denaturing polyacrylamide gel, as described in *Protocol 11*. If there still remains unincorporated [α-^{32}P]UTP in the preparation this can be removed by ethanol precipitation as described in *Protocol 16*.

[a] Which phage RNA polymerase to use will be dictated by the choice of plasmid vector and, in the case of plasmid vectors such as Bluescript or Gemini (which have separate promoters at either end of the polylinker), it will also depend upon the orientation of the cloned probe in the polylinker.
[b] It is also possible to label the probes using other radioactive rNTPs (e.g. [α-^{32}P]GTP) in preference to [α-^{32}P]UTP. In this case the composition of the rNTP mixture should be adjusted appropriately to have a lower final concentration (0.1 mM) of the corresponding radioactive rNTP with the other three rNTPs at 2 mM.

Protocol 15. Northern hybridization of snRNAs using riboprobes

Equipment and reagents

- Electroblot (Hybond nylon membrane) from *Protocol 13*
- Plastic sandwich box with a tight-fitting lid (20 cm × 20 cm in area)
- Formamide, deionized by stirring with a mixed-bed ion-exchange resin — stir 100 ml of formamide with 2 g Amberlite MB-1 for 10 min at room temperature; remove the resin by filtration; store the formamide in aliquots at −20 °C
- 10% SDS
- 0.2 M sodium phosphate buffer, pH 6.5
- Herring sperm DNA (20 mg/ml); boil the solution for 10 min to denature the DNA, cool on ice, and store in aliquots at −20 °C.
- 50 × Denhardt's medium (10 g Ficoll 400, 10 g polyvinylpyrrolidone, 10 g BSA in 1 litre of distilled H_2O)
- 20 × SSC buffer (175.3 g NaCl, 88.2 g trisodium citrate in 1 litre of distilled H_2O). Adjust the pH to 7.0 with 10 M NaOH
- 1 × SSC buffer
- 0.1 × SSC buffer, 0.1% SDS

Continued

4: RNP complexes

Protocol 15. *Continued*

- Hybridization buffer (50% deionized formamide, 5×Denhardts medium, 5×SSC buffer, 20 mM sodium phosphate buffer, pH 6.5, 0.1% SDS, and 200 µg/ml denatured herring sperm DNA). Make this up just before use using stock solutions.
- ^{32}P-labelled probe (see *Protocol 14*)

Method

NB. Wear disposable latex or plastic gloves when handling the electroblot and the radioactive solutions. Use appropriate precautions when working with and disposing of ^{32}P-labelled radioactive materials, including working behind plastic screens.

1. Place the Hybond membrane in the plastic sandwich box and add 100 ml of hybridization buffer (without the probe). Cover the box with its plastic lid and prehybridize for 1–2 h at 37 °C with gentle shaking. Make sure that the membrane is covered with buffer.
2. Add the ^{32}P-labelled riboprobe, allowing approximately 10^4 c.p.m./ml of hybridization buffer. Replace the lid, ensuring that it is tightly closed.
3. Incubate the hybridization reaction at 37 °C for 16–20 h with gentle shaking. *Make sure that the box is clearly marked as radioactive and placed behind adequate shielding.*
4. After hybridization, remove the membrane and carefully and appropriately dispose of the radioactive liquid.
5. Wash the membrane to remove unhybridized probe. To do this, first wash the membrane three times with 1×SSC buffer, each wash comprising 200 ml of buffer for 5 min at room temperature. Then wash once with 200 ml of 0.1×SSC buffer, 0.1% SDS for 30 min at 65 °C.
6. Place the membrane on paper (e.g. 3 MM, Whatman) to dry it, then cover it in plastic wrap, and autoradiograph it at −70 °C using X-ray film in a cassette.

8. Chemical modification interference analysis of RNA

8.1 Strategy

The recognition of specific RNA sequences by RNA-binding proteins and modifying enzymes is a crucial part of RNA processing reactions. It is, therefore, of obvious importance to define the critical nucleotides which are recognized within an RNA motif by sequence-specific RNA-binding proteins. Chemical modification interference is a powerful method for identifying essential bases in an RNA recognition or binding site.

In this technique, RNA is chemically modified at either purine or pyrimidine bases, so that, on average, one base per RNA molecule is modified. Diethyl pyrocarbonate is the reagent used to carboxyethylate purine bases, while hydrazine is used to modify specifically the pyrimidines. In each case the ribose–phosphate chain of the modified RNA becomes susceptible to subsequent cleavage by aniline at the site of modification.

In outline, the method is applied as follows:

(a) A preparation of end-labelled RNA is first chemically modified, part of the sample at purine bases and another part at pyrimidine bases.

(b) The separate populations of RNA molecules, modified at purine or pyrimidine bases, are then used as substrates for either protein-binding or RNA processing; part of each sample is also retained as a control.

Since most RNA processing or binding reactions *in vitro* do not go to completion, it is essential specifically to separate and isolate the reacted or bound RNA at the end of the reaction. The reacted RNAs which are recovered are then cleaved with aniline and fractionated on a denaturing polyacrylamide gel (as in *Protocol 11*). A comparison of the electrophoretic ladders of cleavage products generated by aniline cleavage of the control RNA and RNA recovered from the binding or processing assay will show gaps where bands corresponding to specific nucleotides are missing in the bound or processed RNA sample. The missing bands represent those critical nucleotides where modification inhibited the binding or processing reaction. Conversely, the presence of cleavage bands in the bound or processed RNA implies that modification at these bases does not strictly prevent the binding or processing reaction from taking place.

This modification interference protocol can be used, in principle, with any RNA. It is particularly convenient, however, to apply it to highly-defined substrates made by either *in vitro* transcription (see Chapter 1, Section 2.2) or chemical synthesis (this chapter, Section 2). The method used to isolate the bound or processed RNA at the end of the reaction will depend upon the specific type of assay under study. In many cases this can be done using a native gel system, where the binding of a protein or RNP complex to the RNA substrate causes a 'gel shift' which retards the mobility of the bound RNA. A method for native gel electrophoresis of RNPs is given in *Protocol 12*.

8.2 3' End-labelling of RNA using RNA ligase

The chemical modification reactions are carried out using end-labelled RNA substrates. A method for 3' end-labelling RNA using RNA ligase is given in *Protocol 16*, since 3' end-labelled RNAs allow a more detailed analysis of the involvement of pyrimidine bases in binding or processing than do 5' end-labelled substrates. Alternative procedures for end-labelling RNA are also given in Chapter 2, Section 3.2.

The procedure in *Protocol 16* should yield a labelled RNA product with a specific radioactivity $>10^6$ c.p.m./pmol. However, the labelling efficiency is about 10-fold lower for substrates which have a UMP residue at their 3' terminus. It is important, therefore, to perform the labelling reaction on RNAs terminating in either A, G, or C if at all possible. The RNA must also have a free 3' hydroxyl group.

Protocol 16. 3' End-labelling of RNA substrates using RNA ligase

Reagents

- Purified RNA substrate; this may be either a chemically-synthesized oligoribonucleotide (see *Protocol 1* and Section 2.3) or an RNA made by 'run-off' transcription (Chapter 1, *Protocol 3*)
- T4 RNA ligase
- [5'-^{32}P]pCp (3000 Ci/mmol; Amersham)
- 10 × ligase buffer (0.5 M Tris–HCl, pH 7.9, 0.15 M $MgCl_2$, 30 mM DTT)
- 0.25 mM ATP
- 0.1 mg/ml BSA
- DMSO
- 5.0 M ammonium acetate, pH 7.0
- Glycogen (20 mg/ml)

Method

1. If the concentration of the RNA substrate is less than 5 pmol/µl, re-precipitate the RNA by adding 0.1 vol. of 5 M ammonium acetate, pH 7.0, 0.1 vol. of glycogen (20 mg/ml) and 3 vol. of absolute ethanol. Mix and chill at either −20 °C for at least 3 h or for 1 h in a dry ice/ethanol bath. Centrifuge for 5 min in a microcentrifuge at 10 000 r.p.m. (8800 g) to recover the RNA precipitate. Wash the RNA twice with 70% ethanol (0.5 ml each time) and briefly air dry it. Redissolve the RNA in H_2O at a concentration of 5 pmol/µl. If the original RNA concentration was greater than 5 pmol/µl dilute it in H_2O to a final concentration of 5 pmol/µl.

2. Make up the ligase reaction by mixing together in a microcentrifuge tube at 4 °C:
   - RNA substrate   2 µl (or volume containing 10 pmol)
   - 0.25 mM ATP   2 µl
   - 0.1 mg/ml BSA   2 µl
   - 10 × ligase buffer   2 µl
   - DMSO   2 µl
   - T4 RNA ligase   volume containing 10–20 units
   - sterile H_2O to 20 µl final vol

3. Incubate at 4 °C, either overnight or for at least 10–12 h.

4. Purify the labelled RNA by electrophoresis on a denaturing polyacrylamide gel (see *Protocol 11* and Chapter 1, *Protocol 8*). The procedure described here should provide enough labelled RNA for approximately 10 modification experiments.

8.3 Chemical modification of RNA at purine and pyrimidine bases

A method for chemical modification of RNAs, based on published procedures (29–31), is provided in *Protocol 17*. This protocol is divided into two parts:

- The first part describes the modification at purine bases. The reaction described specifically modifies purines but preferentially modifies A over G by about 4-fold.
- The second part describes the modification at pyrimidines. It can be performed under conditions which modify either (a) U only (b) C only, or (c) U+C. The choice will depend primarily on the sequence of the RNA being studied. Use 3' end-labelled RNA for the pyrimidine modification reactions since the C only and U+C reactions do not work well with 5' end-labelled substrates (30).

Protocol 17. Chemical modification of RNA at purine and pyrimidine bases

Reagents

- RNA substrate 3' end-labelled with ^{32}P as described in *Protocol 16*
- Hydrazine (anhydrous); keep it in a fume hood and store the bottle in a beaker containing a drying agent (**NB** hydrazine is *very toxic*—mark the bottle clearly to indicate a hazardous chemical)
- Hydrazine/0.5 M NaCl (7.3 mg NaCl in 250 µl of hydrazine)
- Hydrazine/3 M NaCl (44 mg NaCl in 250 µl of hydrazine)
- Reagents for ethanol precipitation of RNA (see *Protocol 16*)
- Diethyl pyrocarbonate (DEPC; use fresh)
- Solution A (50 mM sodium acetate, pH 4.5, 1 mM EDTA)
- 1.0 M sodium acetate, pH 4.5 and 0.3 M sodium acetate, pH 3.8

Method

A. *Initial preparation of the RNA*

1. Use starting material preferably 3' end-labelled RNA (*Protocol 16*). Re-precipitate the RNA as described in *Protocol 16*, step 1.

B. *Modification at purine bases*

1. Redissolve the RNA ($\sim 5 \times 10^5$ c.p.m.) in 0.2 ml of solution A in a 1.5 ml microcentrifuge tube.
2. Add 2 µl of fresh DEPC and vortex briefly.
3. Puncture the lid of the tube with a hypodermic needle. Incubate for 2.5–3 min at 90 °C. The hole in the lid allows the CO_2 which is generated inside the tubes during the incubation to escape.

C. *Modification at pyrimidine bases*

1. Follow (a), (b), or (c):
   (a) *U only*
   Redissolve the RNA ($\sim 5 \times 10^5$ c.p.m.) in 10 µl of distilled H_2O. Add 10 µl of anhydrous hydrazine, vortex briefly, and incubate for 10 min at 4 °C.

Continued

Protocol 17. Continued

(b) *C only*
Redissolve the RNA ($\sim 5 \times 10^5$ c.p.m.) in 20 μl of anhydrous hydrazine/3 M NaCl, vortex briefly, and incubate for 30 min at 4 °C.

(c) *U + C*
Redissolve the RNA ($\sim 5 \times 10^5$ c.p.m.) in 20 μl of anhydrous hydrazine/0.5 M NaCl, vortex briefly, and incubate for 30 min at 4 °C.

D. *Recovery of modified RNA*

1. Add 0.2 ml of 0.3 M sodium acetate, pH 3.8, and 0.75 ml of absolute ethanol to each reaction. Mix and leave on ice for 5 min. Collect the precipitated RNA by centrifuging at 12 000 r.p.m. (12 700 g) in a microcentrifuge for 15–20 min at 4 °C.
2. Redissolve the RNA in 0.2 ml of 0.3 M sodium acetate, pH 3.8, and repeat the precipitation as in step 1.
3. Air dry the RNA pellets. The modified RNA can be used in an RNA processing or protein-binding assay as required (see Chapter 3, Section 4; Chapter 5, Section 5).

8.4 Cleavage of modified RNA with aniline

The final step in the chemical modification analysis involves cleaving the modified RNA (which has been recovered from the processing or binding assay) at the site of modification. A procedure for aniline cleavage of modified RNA (30) is provided in *Protocol 18*. The aniline cleavage is performed on RNA modified at purines or pyrimidines as described in *Protocol 17*. Because most *in vitro* RNA processing and binding assays do not go to completion (i.e. do not result in processing or binding of 100% of the substrate RNA), *it is essential to isolate and carry out the aniline cleavage specifically on reacted RNAs*. As a control, modified RNAs that have not been reacted should be cleaved with aniline and analysed in parallel.

Protocol 18. Cleavage of modified RNA with aniline

Reagents

- Purine- or pyrimidine-modified, ^{32}P end-labelled RNA isolated from an RNA processing or binding assay
- Control purine- or pyrimidine-modified, ^{32}P end-labelled RNA that has not been used for a processing or binding assay
- 1.0 M aniline solution: prepare by mixing 50 μl of pure (11 M) aniline^a with 500 μl of 0.3 M sodium acetate, pH 3.8.
- Butan-1-ol
- 1% SDS
- Loading buffer (20 mM Tris–HCl, pH 7.9, 8.0 M urea, 1 mM EDTA, 0.05% bromophenol blue, 0.05% xylene cyanol)

Continued

Protocol 18. *Continued*

Method

1. Ethanol precipitate, as described in *Protocol 16*, step 1, both the modified RNA (purified from the RNA processing or binding assay) and the control modified RNA (that has not been used in a processing or binding assay).
2. Redissolve each RNA in 20 µl of 1 M aniline solution in a 1.5 ml microcentrifuge tube.
3. Incubate for 20 min at 60 °C in the dark.
4. Add 1.3 ml (i.e. fill the tubes) of butan-1-ol and vortex for 30 sec. This stops the reaction and removes the aniline.
5. Centrifuge for 10 min at 12 000 r.p.m. (12 700 g) in a microcentrifuge to pellet the RNA. Discard the supernatant.
6. Add 150 µl of 1% SDS and vortex for 30 sec.
7. Refill the tubes with butan-1-ol and vortex for 30 sec. Spin down the RNA pellets by centrifuging at 12 000 r.p.m. (12 700 g) in a microcentrifuge.
8. Wash the pellets with 1 ml of absolute ethanol.
9. Carefully dry the pellets. Make sure they are completely dry in order to avoid subsequent problems during electrophoresis (a pellet that is not completely dried can give rise to smeared bands on a denaturing gel).
10. Redissolve each pellet in 5 µl of loading buffer and then boil the samples for 2 min.
11. Load the samples on to a denaturing polyacrylamide gel as described in *Protocol 11*. Take care to load approximately equal c.p.m. in each lane. However, in the control lane of unreacted RNA note that there is likely to be a larger number of RNA bands since no bands will be missing due to the interference effect. Therefore, to ensure approximately equal intensities of the bands in the reacted RNA and control lanes, it may be necessary to load more c.p.m. in the control lane.

[a] Redistil the aniline once under nitrogen and store it in the dark at −20 °C.

References

1. Usman, N., Ogilvie, K. K., Jiang, M.-Y., and Cedergren, R. L. (1987). *J. Am. Chem. Soc.*, **109**, 7845.
2. Scaringe, S. A., Francklyn, C., and Usman, N. (1990). *Nucleic Acids Res.*, **18**, 5433.
3. Reese, C. B. and Thompson, E. A. (1988). *J. Chem. Soc. Perkin Trans. 1*, 2881.
4. Sproat, B. S. and Lamond, A. I. (1991). In *Oligonucleotides and analogues: a practical approach* (ed. F. Eckstein), pp. 49–86. IRL Press, Oxford.
5. Pieles, U., Sproat, B. S., and Lamm, G. M. (1990). *Nucleic Acids Res.*, **18**, 4355.
6. Sproat, B. S., Lamond, A. I., Guimil Garcia, R., Beijer, B., Pieles, U., Douglas, M., Bohmann, K., Carmo-Fonseca, M., Weston, S., and O'Loughlin, S. (1991). *Nucleic Acids Res. Symposium Series*, **No. 24**, 59.
7. Polushin, N. N., Pashkova, I. N., and Efimov, V. A. (1991). *Nucleic Acids Res. Symposium Series*, **No. 24**, 49.

8. Iribarren, A. M., Sproat, B. S., Neuner, P., Sulston, I., Ryder, U., and Lamond, A. I. (1990). *Proc. Natl Acad. Sci. USA*, **87**, 7747.
9. Sproat, B. S., Iribarren, A. M., Guimil Garcia, R., and Beijer, B. (1991). *Nucleic Acids Res.*, **19**, 733.
10. Sproat, B. S., Lamond, A. I., Beijer, B., Neuner, P., and Ryder, U. (1989) *Nucleic Acids Res.*, **17**, 3373.
11. Barabino, S., Sproat, B. S., Ryder, U., Blencowe, B. J., and Lamond, A. I. (1989). *EMBO J.*, **8**, 4171.
12. Lamond, A. I., Sproat, B., Ryder, U., and Hamm, J. (1989) *Cell*, **58**, 383.
13. Blencowe, B. J., Sproat, B. S., Ryder, U., Barabino, S., and Lamond, A. I. (1989). *Cell*, **59**, 531.
14. Lamm, G. M., Blencowe, B. J., Sproat, B. S., Iribarren, A., Ryder, U., and Lamond, A. I. (1991). *Nucleic Acids Res.*, **19**, 3193.
15. Barabino, S. M. L., Blencowe, B. J., Ryder, U., Sproat, B. S., and Lamond, A. I. (1990). *Cell*, **63**, 293.
16. Wassarman, D. A. and Steitz, J. A. (1991). *Mol. Cell. Biol.*, **11**, 3432.
17. Gröning, K., Palfi, Z., Gupta, S., Cross, M., Wolff, T., and Bindereif, A. (1991). *Mol. Cell. Biol.*, **11**, 2026.
18. Palfi, P., Günzl, A., Cross, M., and Bindereif, A. (1991). *Proc. Natl Acad. Sci. USA*, **88**, 9097.
19. Smith, H. O., Tabiti, K., Schaffner, G., Soldati, D., Albrecht, U., and Birnstiel, M. L. (1991). *Proc. Natl Acad. Sci. USA*, **88**, 9784.
20. Blencowe, B. J. (1991). PhD Thesis, University of London.
21. Hames, B. D. (1990). One-dimensional polyacrylamide gel electrophoresis. In *Gel electrophoresis of proteins: a practical approach* (ed. B. D. Hames and D. Rickwood), pp. 1-148. IRL Press, Oxford.
22. Krämer, A., Keller, W., Appel, B., and Lührmann, R. (1984). *Cell*, **38**, 299.
23. Krainer, A. R. and Maniatis, T. (1985). *Cell*, **42**, 725.
24. Black, D. L., Chabot, B., and Steitz, J. A. (1985). *Cell*, **42**, 737.
25. Berget, S. M. and Robberson, B. L. (1986). *Cell*, **46**, 691.
26. Black, D. L. and Steitz, J. A. (1986). *Cell*, **46**, 697.
27. Pan, Z-Q. and Prives, C. (1988). *Science*, **241**, 1328.
28. Hamm, J., Dathan, N. A., and Mattaj, I. (1989). *Cell*, **59**, 159.
29. Peattie, D. A. (1979). *Proc. Natl Acad. Sci. USA*, **76**, 1760.
30. Conway, L. and Wickens, M. (1987). *EMBO J.*, **6**, 4177.
31. Rymond, B. C. and Rosbash, M. (1988). *Genes and Dev.*, **2**, 428.

5

Analysis of ribonucleoprotein interactions

CINDY L. WILL, BERTHOLD KASTNER,
and REINHARD LÜHRMANN

1. Introduction

Elucidation of the precise function of small ribonucleoprotein particles (snRNPs) in RNA processing will be greatly aided by information regarding their biochemical composition and higher order structure. To date, significant biochemical and structural information has been gathered for the mammalian spliceosomal snRNPs U1, U2, U5, and U4/U6, although their characterization is far from complete. The snRNA component of these functionally important RNP particles has been shown to be evolutionarily conserved and, with the exception of U6, possesses a unique 5' cap structure, namely 2,2,7-trimethylguanosine (m_3G). The spliceosomal snRNPs share a set of eight common proteins (referred to as B, B', D1, D2, D3, E, F, and G) which are tightly associated with the snRNA component of the snRNP (1). Particle-specific proteins are also associated with several snRNP species (U1, U2, U5, and the tri-snRNP [U4/U6.U5] complex), although in many cases their interaction is highly dependent on the ionic strength of the particle's environment (1). For this reason, the sedimentation coefficient of an individual snRNP may vary. At intermediate salt concentrations (approximately 0.3–0.5 M), U1, U2, and U4/U6 sediment as 12S particles and U5 as a 20S particle, whereas at lower ionic strengths, a 17S U2 and a 25S [U4/U6.U5] tri-snRNP particle are also observed. Investigation of intermolecular interactions within RNP particles can be accomplished by a variety of procedures. Since most structural studies are facilitated by the analysis of individual RNP complexes in the absence of contaminating macromolecules, this chapter starts by describing the approaches commonly used for the isolation of snRNP particles. Several procedures are then described which are particularly useful for analysing RNA–protein interactions and/or RNP higher order structure. These include snRNP reconstitution, immunoprecipitation, RNA–protein cross-linking, nuclease protection, and immunoelectron microscopy.

5: *Analysis of RNP interactions*

2. Immunoaffinity purification of RNP complexes

2.1 Strategy

An important approach to the purification of snRNPs involves the use of antibodies directed against either an individual protein or a group of proteins or against the 5' cap (m_3G) of the RNA component of the particle. Immunoaffinity chromatography has been used successfully to purify snRNP particles from other nuclear constituents and to separate distinct snRNP species from one another. This procedure entails the covalent attachment of anti-snRNP antibodies to a solid-phase matrix. RNP complexes are subsequently bound by the immobilized antibodies, the matrix is washed to remove unbound contaminating material, and the adsorbed RNP complexes are eluted in pure form by disruption of the antigen–antibody interaction. This interaction can be disrupted in a number of ways, ranging from non-specific elution with strongly denaturing reagents to specific elution with an excess of the purified antigen comprising the cognate epitope, a method which also preserves the snRNP in its native state. Although RNP particles are obviously disrupted if strong denaturants are used, purified snRNA or snRNP proteins can still be obtained (2). With this method of elution, snRNP-specific antibodies from any source may be used (for instance, sera from autoimmune patients, polyclonal sera, or supernatants from monoclonal hybridomas). If, however, snRNPs are to be analysed structurally, or even functionally, elution must be performed under very mild conditions in order to avoid altering their native structure. This is best accomplished using monoclonal or affinity-purified polyclonal antibodies whose epitope is well defined; for example, those which have been raised against a synthetic peptide or modified nucleoside (i.e. m_3G or m^6A). The use of such antibodies allows native snRNP particles to be eluted by competition with an excess of the cognate peptide or nucleoside (see Section 2.3 for examples). There is an additional advantage with this procedure, the resultant peptide–antibody interaction can often be disrupted under moderately harsh conditions without significant loss of antibody activity. Thus, the immunoaffinity columns can often be used repeatedly.

2.2 Immobilization of antibodies on Sepharose matrices

The matrix of choice for immunoaffinity chromatography is Protein A–Sepharose (PAS). Aside from the added advantages of being relatively inexpensive and easy to work with, Protein A binds the Fc region of an IgG molecule and, therefore, orientates the antibody in a favourable position for antigen interaction. This is not the case with most chemically-activated matrices. Care must be taken when working with monoclonal antibodies since different IgG subclasses, as well as IgGs from different species, possess variable affinities for Protein A (see ref. 3 for a list of relative affinities). However, a poor affinity for PAS can often be circumvented by using Protein G–Sepharose which has a different IgG binding spectrum, or, alternatively, by using a chemically-activated matrix such as CNBR-activated Sepharose 4B. If activated matrices are employed, purified antibodies (for instance, by gel filtration) should be used to prevent the simultaneous binding of proteases and RNases.

Protocol 1, adapted from Harlow and Lane (3), describes the procedure recommended by the authors for preparing the solid support for immunoaffinity purifications. The IgG is first bound to PAS and then the bound IgG is attached covalently using dimethylpimelimidate to provide a more stable immunoadsorbent.

Protocol 1. Immobilization of antibodies on Protein A–Sepharose

Equipment and reagents

- Protein A–Sepharose CL-4B (Pharmacia)[a]
- End-over-end rotating stirrer
- PBS (20 mM potassium phosphate, pH 8.0, 0.13 M NaCl); with and without 0.02% NaN_3
- Antibody preparation, e.g. antiserum, hybridoma supernatant or ascites fluid, purified IgG
- 1.0 M Tris–HCl, pH 8.0
- Dimethylpimelimidate (Pierce; store in a dessicator at 4 °C)
- 0.2 M sodium borate buffer, pH 9.0
- 0.2 M sodium borate buffer, pH 9.0, containing 20 mM dimethylpimelimidate prepared directly before use
- 0.2 M ethanolamine–HCl, pH 8.0
- Protein sample buffer (60 mM Tris–HCl, pH 6.9, 1 mM EDTA, 16% glycerol, 2% SDS, 50 mM DTE, 0.1% bromophenol blue). Note that the DTE should be added shortly before use.
- Chromatography column sterilized by autoclaving (length and diameter are dictated by the matrix bed volume)

Method

A. *Immobilization of antibodies*

1. Swell Protein A–Sepharose (PAS) in 5 vol. of PBS. Stir the slurry overnight by end-over-end rotation at 4 °C.
2. Pellet the PAS by centrifugation at 10 000 g for 10 sec and wash it three times with 5 vol. of PBS.
3. Adjust the pH of the antibody preparation to 8.0 by the addition of 0.1 vol. of 1.0 M Tris–HCl, pH 8.0.
4. Pellet the washed PAS as in step 2 and remove the PBS. Combine the PAS and antibody and mix the slurry on the rotary stirrer at room temperature for 1 h. PAS will bind ~10–20 mg of IgG per millilitre packed volume. The approximate concentrations of IgG in serum, ascites fluid and hybridoma supernatant are 10–15 mg, 1–10 mg, and 20–50 µg per ml, respectively.
5. Pellet the PAS as in step 2 and discard the supernatant.
6. Wash the PAS with 5 vol. of 0.2 M sodium borate, pH 9.0, and pellet it as in step 2.

Continued

Protocol 1. *Continued*

7. Repeat the washing (step 6) three times. Before pelleting the PAS for the last time, remove and retain a sample of the slurry equivalent to 10 µl (packed volume) of the matrix for later analysis of the amount of IgG bound.
8. Attach the antibody covalently to the PAS by adding 10 vol. of 0.2 M sodium borate, pH 9.0, containing 20 mM dimethylpimelmidate[b]. Mix the slurry by rotation at room temperature for 1 h. Remove another sample of the matrix as in step 7.
9. Stop the coupling reaction by pelleting the PAS and washing it twice (30 min each) with 10 vol. of 0.2 M ethanolamine–HCl, pH 8.0.
10. Wash the PAS four times with 5 vol. of PBS and transfer the washed matrix to a sterile column.
11. Equilibrate the column with PBS containing 0.02% NaN_3 and store it at 4 °C.

B. *Monitoring the efficiency of IgG linkage*

1. Vacuum dessicate the matrix samples taken before and after coupling (steps 7 and 8).
2. Resuspend the PAS of each sample in 25 µl of protein sample buffer by vortexing for 5 min and then boiling for 5 min.
3. Analyse the samples by SDS–PAGE as described in *Protocol 7*. The absence of an IgG heavy chain band (M_r 55 000) in the samples taken after coupling indicates efficient linkage of the IgG to the Sepharose.

[a] Protein G–Sepharose may be substituted for the PAS for antibodies possessing a low affinity for PAS.
[b] Efficient coupling requires a final pH above 8.3.

Monoclonal antibodies with a reduced affinity for both Protein A and Protein G may be coupled indirectly through an anti-immunoglobulin antibody. For instance, monoclonal antibodies of the IgM class can be attached to the PAS via an anti-IgM IgG intermediate. *Protocol 2* describes a modification of the procedure in *Protocol 1* for such indirect coupling.

Protocol 2. Indirect coupling of monoclonal antibodies to Sepharose

Equipment and reagents

- Reagents and equipment listed in *Protocol 1*
- Anti-immunoglobulin antibody (affinity-isolated IgG from Sigma, dissolved in PBS)
- 0.1 M Tris–HCl, pH 8.0

Method

1. Prepare the PAS as described in *Protocol 1*, steps 1 and 2.

Continued

Protocol 2. *Continued*

2. Mix the washed PAS with the anti-immunoglobulin antibody (~ 2 mg/ml packed volume of matrix) and mix the slurry on the rotary stirrer at room temperature for 1 h.
3. Pellet the matrix and discard the supernatant.
4. Wash the matrix five times using 2 vol. of 0.1 M Tris–HCl, pH 8.0, each time.
5. Resuspend the matrix in an excess of the specific second antibody. The amount of antibody which will bind depends on the amount of primary antibody bound to the PAS in step 2. One milligram of affinity-isolated anti-IgM IgG will bind 1.92 mg of mouse IgM.
6. Mix the slurry on the rotary stirrer for 2 h at room temperature.
7. Complete the column preparation as described in *Protocol 1*, steps 5–11.
8. Monitor the efficiency of antibody linkage to the PAS as described in *Protocol 1B*.

2.3 Immunoaffinity chromatography of snRNPs

This section describes two immunoaffinity procedures for the isolation of snRNPs from a nuclear extract (4, 5) prepared from HeLa cells by the method of Dignam *et al.* (6) and as described in Chapter 3 (*Protocol 6*). In the first procedure, U1, U2, U4/U6, and U5 snRNPs are isolated from the nuclear extract with an antibody against the m_3G cap. In the second method, U1 snRNPs or [U4/U6.U5] tri-snRNP complexes are isolated from the nuclear extract under low salt conditions with a monoclonal antibody (H386) which reacts with a linear epitope in the C-terminal region of the U1-specific 70 kDa protein. In this latter instance, the extract is initially subjected to centrifugation on glycerol gradients to separate the 12S U1 snRNPs from both the 20S U5 and 25S [U4/U6.U5] complexes. The H386 antibody cross-reacts with the U5-specific 100 kDa protein. Thus, it can be used to isolate complexes containing either U1 or U5, depending on which fractions from the glycerol gradients are used for immunoaffinity chromatography. Bound snRNP particles are eluted specifically under native conditions with either an excess of the nucleoside (see *Protocol 3*) or the peptide (see *Protocol 4*) comprising the cognate epitope. Note that m^7G nucleoside is used for elution in *Protocol 3* since it cross-reacts with the anti-m_3G antibodies.

Protocol 3. Immunoaffinity chromatography of snRNPs using an antibody against a modified ribonucleoside

Equipment and reagents

- Nuclear extract from HeLa cells (5 mg protein/ml in 20 mM Hepes–KOH, pH 7.9, 25% glycerol, 0.42 M NaCl, 1.5 mM $MgCl_2$, 0.2 mM EDTA, 0.5 mM DTE, and 0.5 mM PMSF); prepared as described in Chapter 3 (*Protocol 6*)

Continued

Protocol 3. Continued

- C-buffer (20 mM Hepes–KOH, pH 7.9, 0.25 M NaCl, 1.5 mM $MgCl_2$, 0.2 mM EDTA, 0.5 mM DTE (Merck), and 0.5 mM PMSF (Serva))
- C-5 buffer (C-buffer containing 5% glycerol)
- C-5 buffer containing 6 M urea
- Immunoaffinity column prepared as described in *Protocol 1* (1 ml packed volume of PAS containing 20 mg of IgG from rabbit anti-m_3G serum, R1131[a], equilibrated in C-5 buffer)
- Beckman ultracentrifuge, Ti70 rotor and tubes (or equivalent)
- 5 μm and 1.2 μm disposable membrane filters (Schleicher and Schuell)
- m^7G nucleoside (Sigma; 20 mM in C-5 buffer)

Method

NB *Perform all steps at 4 °C*

1. Centrifuge the nuclear extract at 165 000 g for 30 min in the Beckman Ti70 rotor to remove large aggregates.
2. Use a syringe to pass the supernatant through a 5 μm membrane filter, followed by a 1.2 μm filter. Dilute the extract with 0.5 vol. of C-buffer.
3. Load the diluted extract, at a flow rate of 6.6 ml/h, on to the anti-m_3G column equilibrated with C-5 buffer. A 1 ml column should be loaded with 15 ml of extract (equivalent to 3.3×10^9 cells).
4. Wash the column with 5 bed vol. of C-5 buffer at a flow rate of 30 ml/h.
5. Elute the snRNPs with one bed vol. of m^7G nucleoside (20 mM) at a flow rate of 12 ml/h, collecting 0.1 bed vol. fractions. After two-thirds of the m^7G solution has penetrated the column, stop the flow for 30 min. Then allow the remaining m^7G solution to enter the column. Complete the snRNP elution with 1.5 bed vol. of C-5 buffer.
6. Determine the protein concentration of each fraction (ref. 7) and analyse the fractions for snRNP protein and RNA as described in *Protocols 6–8*.
7. Freeze those fractions containing the purified spliceosomal snRNPs in liquid nitrogen and store them at −70 °C.
8. Regenerate the column by washing it with 4 bed vol. of C-5 buffer containing 6 M urea, followed by 10 bed vol. of C-5 buffer at a flow rate of 30 ml/h.

[a] An alternative is to use a column made using CNBr–activated Sepharose 4B (Pharmacia) linked to H-20 anti-m_3G monoclonal antibody (6 mg antibody/ml Sepharose) using the procedure recommended by the manufacturer.

Under the elution conditions described in *Protocol 3* (i.e. 0.25 M NaCl), significant amounts of 25S [U4/U6.U5] tri-snRNP complexes are detected in the eluate. If the NaCl concentration of the elution buffer is increased to 0.42 M, U5 and U4/U6 snRNPs will be recovered as 20S and 12S complexes, respectively. Hence, the ionic strength of the wash and elution buffer may be varied to yield snRNP particles containing various protein subsets and, in this way, tightly- and loosely-associated proteins can readily be distinguished. *Figure 1* illustrates the immunoaffinity purification of snRNPs from HeLa cell nuclear extract using an anti-m_3G antibody. *Protocol 4* describes the immunoaffinity chromatography of 12S U1 snRNPs.

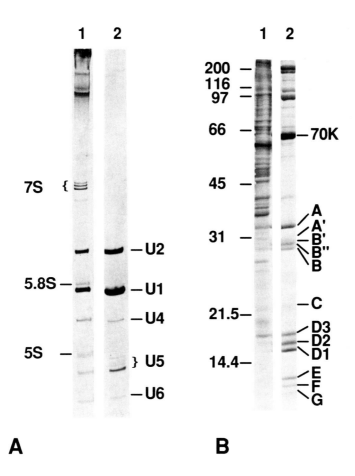

Figure 1. One-step purification of spliceosomal snRNPs by anti-m$_3$G affinity chromatography. Immunoaffinity chromatography was performed as described in *Protocol 3*. A. RNA from HeLa cell nuclear extract (lane 1) or anti-m$_3$G affinity-purified snRNPs (lane 2) was isolated by phenol extraction and ethanol precipitation (see *Protocol 6*). In each case 1.0 µg of RNA was fractionated on a 10% denaturing polyacrylamide/urea gel and silver stained (see *Protocol 8*). The positions of snRNAs are shown on the right and marker RNAs on the left. B. Proteins from the nuclear extract (lane 1) or affinity-purified snRNPs (lane 2) were isolated by phenol extraction and acetone precipitation (see *Protocol 6*). In each case 10 µg samples were fractionated on a 12% polyacrylamide–SDS gel and stained with Coomassie brilliant blue (see *Protocol 7*). The positions and M_r (kDa) of marker proteins are indicated on the left and a subset of the snRNP proteins is identified on the right.

5: Analysis of RNP interactions

Protocol 4. Immunoaffinity chromatography of snRNPs using an antibody against a defined peptide epitope

Equipment and reagents

- Nuclear extract from HeLa cells (see *Protocol 3*)
- Immunoaffinity column: PAS coupled to H386 monoclonal IgM against the U1 snRNP 70K protein through a goat anti-IgM intermediate as described in *Protocol 2*
- G-buffer (20 mM Hepes–KOH, pH 7.9, 0.15 M KCl, 1.5 mM $MgCl_2$, 0.2 mM EDTA, 0.5 mM DTE, 0.5 mM PMSF, 5% glycerol)
- G-buffer containing 10% and 30% glycerol
- Competing peptide: synthesized on an automated peptide synthesizer (e.g. Applied Biosystems 430A peptide synthesizer) and purified by gel filtration chromatography on a Sephadex G-25 column (for immunoaffinity preparations using the H386 antibody, use a 32mer with the sequence DRDRERRRSHRSER-ERRRDRD RDRDRDREHKR)
- Beckman ultracentrifuge, SW40 Ti rotor, and tubes (or equivalent)
- Gradient former and gradient fractionator with peristaltic pump and UV detector
- 10 mM sodium phosphate buffer, pH 7.2, with and without 3.5 M $MgCl_2$

Method

NB *Carry out all steps at 4 °C.*

1. Dialyse the nuclear extract for 2.5 h against 30 vol. of G-buffer. Change the buffer and dialyse for another 2.5 h.
2. Use a gradient former to prepare 11 ml glycerol gradients (10%–30% glycerol in G-buffer) in SW40 tubes.
3. Load 2 ml of nuclear extract on to each glycerol gradient and centrifuge the gradients at 150 000 g for 18 h to separate the 12S U1 snRNPs from the 20S U5 and 25S [U4/U6.U5] tri-snRNP complexes.
4. Fractionate the gradients by upward displacement collecting 26 × 0.5 ml fractions.
5. Monitor the A_{280} of the fractions; 12S and 25S complexes are usually found in fractions 10–14 and 19–22, respectively. To confirm the composition of each fraction, analyse the protein and RNA content of 50 µl of each fraction by PAGE as described in *Protocols 6–8*.
6. Equilibrate the immunoaffinity column by washing with 10 bed vol. of G-buffer.
7. Pool the gradient fractions containing the U1 snRNPs[a] and apply the material to the immunoaffinity column (1 ml packed vol. per 3.5 ml of snRNP preparation).
8. Recirculate the preparation (using a pump) through the column for 2 h.
9. Wash the column with 20 bed vol. of G-buffer.
10. Elute the bound snRNPs with 2 bed vol. of G-buffer containing 10 µM of the competing peptide[b]. Collect 15 fractions (each of 0.15 bed vol.).

Continued

Protocol 4. *Continued*

11. Determine the protein and RNA composition of one-fifth of each fraction by performing PAGE as described in *Protocols 6–8*. Eluted U snRNPs are typically first detected in fraction 5 and reach a peak in fractions 8–10.
12. Freeze those fractions containing U1 snRNPs in liquid nitrogen and store them at −70 °C.
13. Regenerate the column by washing with 10 vol. of 10 mM sodium phosphate buffer, pH 7.2, followed by 5 vol. of buffer containing 10 mM sodium phosphate buffer, pH 7.2 and 3.5 M $MgCl_2$, and, finally, with 10 vol. of G-buffer.

[a] Alternatively, those gradient fractions containing 20S U5 or 25S [U4/U6.U5] snRNPs may be applied.
[b] The concentration of competing peptide required to displace the bound snRNP must be determined in a pilot experiment for each peptide.

3. Ion-exchange chromatography of snRNPs

The fractionation of the individual snRNP species present in a nuclear extract (8) or in affinity-selected snRNP preparations can also be accomplished by ion-exchange chromatography on Mono Q columns under FPLC conditions (4). Mono Q column chromatography can also be used for the isolation of subpopulations of a particular snRNP species (i.e. U1 snRNPs which lack one or more specific proteins). Potentially, protein-deficient particles can be used to investigate the interaction of the missing protein with its cognate RNP particle and also its function in the splicing reaction. For example, U1 snRNPs lacking the particle-specific C protein have been used to study the interaction of ^{35}S-labelled C protein, synthesized by translation *in vitro*, with the U1 snRNP particle. Through these studies, those regions of the C protein necessary for binding to U1 snRNP have been elucidated (9). Although ion-exchange chromatography on Mono Q provides a straightforward approach for the separation of individual snRNP particles, only snRNPs possessing relatively tightly-associated polypeptides can be obtained, due to the relatively high ionic strength of the elution buffers employed and the strength of the ionic exchange resin.

Protocol 5 describes the purification of snRNPs on Mono Q. Using this procedure, the majority of the 12S U1 sRNPs elute in the first peak of the gradient at about 0.37 M KCl. This peak also contains a significant amount of 20S U5 snRNP but, if necessary, this can be removed by glycerol gradient centrifugation either before or after the Mono Q chromatography (see *Protocol 4*, steps 2–5). Centrifugation of the U1/U5-containing fraction from Mono Q chromatography must be carried out before freezing the purified samples since the high molecular weight U5 proteins tend to dissociate, forming 12S U5 particles which will co-sediment with the 12S U1 snRNPs during gradient

5: Analysis of RNP interactions

centrifugation. A second U5 snRNP peak elutes from the Mono Q column at 0.48 M KCl, whereas the majority of U2 and U4/U6 snRNPs elute at 0.49 M and 0.55 M KCl, respectively.

Protocol 5. Chromatography of snRNPs on Mono Q

Equipment and reagents

- FPLC apparatus and Mono Q column (1 ml bed vol; Pharmacia)
- Mono Q buffer (20 mM Tris–HCl, pH 7.0, 1.5 mM $MgCl_2$, 0.5 mM DTE, 0.5 mM PMSF, 2 μg/ml leupeptin)
- Mono Q buffer containing 50 mM KCl and 1.0 M KCl
- Spliceosomal snRNPs (10 mg) prepared as described in Protocols 3 or 4

Method

NB *Carry out all steps at 4 °C*

1. Dilute the spliceosomal snRNP preparation with Mono Q buffer until the final concentration of monovalent cations is less than 0.2 M.
2. Equilibrate the Mono Q column in Mono Q buffer containing 50 mM KCl.
3. If necessary, centrifuge the snRNP preparation on glycerol gradients (see *Protocol 4*, steps 2–5).
4. Load the snRNP preparation on to the Mono Q column at a flow rate of 2 ml/min (pressure 3 MPa).
5. Wash the column with Mono Q buffer containing 50 mM KCl until the $A_{280} = 0$.
6. Elute the snRNPs with a linear gradient of 50 mM KCl to 1 M KCl in Mono Q buffer (flow rate 1 ml/min). The authors usually run a step gradient with the following increases in Mono Q buffer containing 1 M KCl: 4 min with a 5.4% increase per min, 30 min with a 1% increase per min, and 10 min with a 4.2% increase per min.
7. Collect 1 ml fractions, monitoring the elution of snRNP by measuring the A_{280} of the fractions.
8. Determine the protein concentration of the peak fractions (7) and analyse the RNA and protein content by PAGE (see *Protocols 6–8*).
9. Freeze the purified snRNPs in liquid nitrogen and store them at −70 °C.

Analyses of 12S U1, U2, and U4/U6 particles and 20S U5 particles purified on Mono Q are presented in *Figure 2*. The yield of fractionated snRNPs is generally higher than 50% and concentrated samples (2–3 mg/ml protein) can be obtained. U1 snRNPs lacking one or more of their specific proteins (A, C, 70K) can be isolated by performing steps 1–3 of *Protocol 5* at higher temperatures (10–37 °C). Small amounts of protein-deficient U1 particles are also often observed after chromatography at 4 °C.

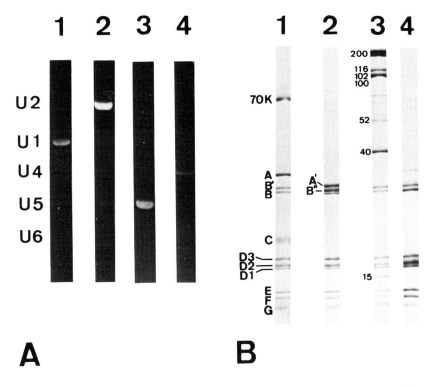

Figure 2. Fractionation of spliceosomal snRNPs by Mono Q chromatography. Mono Q chromatography was performed as described in *Protocol 5*. Lanes 1–4 correspond to column fractions containing U1, U2, U5, and U4/U6 particles, respectively. In each case the isolation and fractionation of the RNA (panel A) and protein constituents (panel B) of the purified snRNPs were performed as described in the legend to *Figure 1*, except that the RNA was visualized by staining with ethidium bromide. Common and particle-specific proteins are identified (panel B, at left of lanes 1–3). The authors thank Thomas Lehmeier for providing this figure.

4. Analysis of snRNP protein and RNA by gel electrophoresis

PAGE separation of proteins and RNA is a much-used technique in the analysis of snRNPs. Prior to gel electrophoresis, the snRNA and protein components of an snRNP particle are separated by extraction with phenol. The various snRNA species are typically separated according to molecular weight on denaturing polyacrylamide/urea gels (see Chapter 1, *Protocol 8*; Chapter 4, *Protocol 11*); snRNP proteins, on the other hand, are separated on polyacrylamide gels in the presence of SDS. SDS–polyacrylamide gels consist of a short, upper stacking gel, where protein samples are concentrated into a small volume, and a lower, resolving gel, where proteins are separated according to

5: Analysis of RNP interactions

to molecular weight. In order to achieve good resolution of all snRNP protein species, in particular D1, D2, and D3, it is essential to use concentrations of TEMED five to seven times higher than that employed in standard SDS–polyacrylamide gels (10). The authors typically separate snRNP proteins on 12% polyacrylamide–SDS gels, 10–20 cm in length.

Protocol 6 describes the procedure for the fractionation of snRNPs into protein and RNA by phenol extraction prior to gel electrophoresis. In *Protocol 7* the procedure recommended by the authors for the electrophoretic separation of snRNP proteins is described. Finally, a silver-staining procedure, adapted from Merril *et al.* (11), which can be used to detect electrophoretically-separated snRNP protein or RNA, is presented in *Protocol 8*.

Protocol 6. Phenol extraction of snRNPs

Reagents

- snRNP sample for analysis (~7.5–10 µg protein are required for subsequent analysis by gel electrophoresis)
- Phenol:chloroform:isoamylalcohol (50:49:1 by vol.) with 0.1% (w/v) 8-hydroxyquinoline, saturated by equilibration with 10 mM Tris–HCl, pH 8.0, 0.5 mM EDTA
- 3 M sodium acetate, pH 5.2
- 10 mg/ml carrier tRNA (*E. coli* type XX from Sigma, phenol:chloroform:isoamyl-alcohol extracted and dissolved in sterile water) or 20 mg/ml glycogen (molecular biology grade from Boehringer Mannheim prepared in sterile water)

Method

1. To the snRNP sample add an equal vol. of the phenol:chloroform:isoamyl-alcohol mixture[a]. Shake vigorously for 5 min on a multisample vibrator or vortex for 1 min.

2. Separate the phases by centrifugation for 5 min at 12000 g in a microcentrifuge.

3. Carefully remove the upper[b] aqueous phase, taking care not to remove the denatured protein normally visible at the interface, and transfer it to a fresh microcentrifuge tube. Retain the lower phenol phase for isolation of proteins (step 6).

4. Add to the aqueous phase (which contains the RNA), 2.5 vol. of absolute ethanol, 0.1 vol. of 3 M sodium acetate[c] (pH 5.2) and, if the concentration of RNA is less than 1 µg/ml, carrier tRNA to a final concentration of 10 µg/ml. If the RNA is subsequently to be end-labelled (see Section 6.3), replace the tRNA with 33 µg/ml of glycogen.

5. Mix the sample by vortexing briefly and allow the RNA to precipitate for 1 h at −70 °C or overnight at −20 °C.

6. Add to the phenol (yellow, lower) phase from step 3, 5 vol. of acetone.

7. Mix the sample and allow the protein to precipitate for 1 h at −70 °C or overnight at −20 °C.

Continued

Protocol 6. *Continued*

8. Recover the precipitated RNA and protein (steps 5 and 7), by centrifuging the mixtures for 10 min at 12 000 g in a microcentrifuge.
9. Carefully remove and discard the supernatant, in each case.
10. Wash the RNA and protein precipitates with 80% ethanol and dry them under vacuum.

[a] To isolate immunoprecipitated RNA (see *Protocol 11*), add 0.15 ml of H_2O to the pelleted Protein A–Sepharose before phenol extraction.
[b] Note that the phenol and aqueous phases may be inverted if the sample contains extremely high levels of salt.
[c] If the samples already contain more than 0.3 M salt (for instance Mono Q fractionated samples), the addition of sodium acetate should be omitted.

Protocol 7. Analysis of snRNP protein by gel electrophoresis

Equipment and reagents

- Power source, vertical electrophoresis chamber, glass plates (15–25 cm in length), 0.5 mm thick spacers and comb
- TEMED (*N,N,N',N'*-tetramethylethylene diamine)
- Protein molecular weight markers (M_r range 14–200 kDa)
- 30% (w/v) polyacrylamide solution with 0.8% bisacrylamide (Roth)
- 10% (w/v) ammonium persulphate (prepared weekly, stored at 4 °C)
- Isobutanol (saturated with water)
- Electrophoresis buffer (25 mM Tris base– 0.192 M glycine, pH 8.3, 0.05% SDS). **Note** that all Tris-containing buffers must be prepared with Tris base (*not* Tris–HCl or Trizma) and, in the subsequent buffers, the pH adjusted with HCl.
- 4 × stacking gel buffer (0.5 M Tris–HCl, pH 6.8, 0.4% SDS)
- 4 × resolving gel buffer (1.5 M Tris–HCl, pH 8.8, 0.4% SDS)
- Protein sample buffer (see *Protocol 1*)
- Plastic or stainless steel container
- Tilting platform shaker
- Coomasie staining solution (50% methanol, 9.2% acetic acid, 0.25% Coomasie brillant blue)
- Destain solution (40% methanol, 10% acetic acid)
- Sterile water

Method

NB *Acrylamide is toxic! Gloves should be worn at all times*

1. Assemble the gel casting sandwich[a]. (Assembly methods vary, depending on the plate type and manufacturer.) Place the gel casting sandwich in a vertical position.
2. Combine the following solutions, adding TEMED last:
    - 4 × resolving gel buffer   7.5 ml
    - 30% polyacrylamide solution   12.0 ml

Continued

Protocol 7. Continued

- sterile water   10.5 ml
- 10% ammonium persulphate   100 µl
- TEMED   100 µl

3. Mix briefly by swirling and pour, or pipette, the acrylamide solution immediately into the gel casting sandwich, leaving sufficient space for the stacking gel (typically 3–4 cm[b]).
4. With a Pasteur pipette, carefully overlay the acrylamide solution with saturated isobutanol.
5. Allow the acrylamide to polymerize.
6. Remove the isobutanol by inverting the glass sandwich, rinse with sterile water, and remove excess liquid with a piece of filter paper.
7. Combine the following solutions, adding TEMED last:
   - 4 × stacking gel buffer   2.5 ml
   - 30% polyacrylamide solution   1.66 ml
   - sterile water   5.84 ml
   - 10% ammonium persulphate   50 µl
   - TEMED   25 µl
8. Mix by swirling briefly and pour immediately into the gel casting sandwich. Place the comb between the glass plates, taking care to avoid trapping air bubbles, and allow the acrylamide to polymerize.
9. Remove the comb, mount the glass sandwich in the gel chamber, and fill the upper and lower chamber with electrophoresis buffer.
10. Add ~10 µl protein sample buffer to the lyophilized protein samples[c] (see *Protocol 6*, step 10) and mix by shaking vigorously for 10 min. Heat the samples for 5 min at 95 °C.
11. Pipette the samples into the wells of the gel. Include a sample of a protein molecular weight marker mixture in another well.
12. Apply a constant current of 15 mA (for a 20 cm wide gel) until the bromophenol blue marker dye reaches the resolving gel, then increase the current to 20–25 mA. Stop the electrophoresis as soon as the marker dye has migrated to the bottom of the gel.
13. Remove the gel from the glass sandwich by prising apart the plates and place it in a plastic or stainless-steel container.
14. Add the Coomasie staining solution until the gel is completely covered and incubate the gel on a tilting platform shaker for 30 min.

---

[a] Plates which allow for a 10–20 cm long resolving gel should be used.
[b] The length of the stacking gel is dependent upon the length of the comb's teeth. A minimum of 1 cm must be left between the base of the teeth (bottom of the wells) and the resolving gel.
[c] 7.5–10 µg of snRNP protein are normally loaded in a 0.5 cm wide well.

Continued

Protocol 7. *Continued*

15. Remove the Coomasie solution and add the destain solution. Destain the gel until the desired staining intensity is reached. This generally requires 3–4 changes of the destain solution.

NB The gel may be dried or stored in 5% acetic acid without significant reduction in staining intensity. If the amount of protein is not sufficient for detection with Coomasie blue, the gel may be silver stained as described in *Protocol 8*.

Protocol 8. Silver staining snRNP protein or RNA

Equipment and reagents

- SDS–polyacrylamide or polyacrylamide/urea gel containing electrophoretically separated snRNP protein or RNA[a], respectively (see *Protocol 7* or Chapter 1, *Protocol 8*); note that only gels <0.75 mM thick are suitable for this staining procedure.
- Tilting platform shaker, plastic or stainless steel container, and sterile water (see *Protocol 7*)
- Fix solution: 40% methanol, 10% acetic acid
- Prestain solution: 10% ethanol, 5% acetic acid
- 12 mM $AgNO_3$
- Developing solution: 0.28 M Na_2CO_3, 0.0185% formaldehyde
- 5% acetic acid

Method

NB *Perform all steps at room temperature and wear gloves at all times*

1. Place the gel in a plastic or stainless-steel container and add fix solution until the gel is completely covered. Place the container on a tilting platform shaker for 30 min. Perform the following steps in an identical manner.
2. Wash for 15 min with prestain solution. Repeat.
3. Remove the prestain solution and wait 5 min (to allow for oxidation) before washing briefly with sterile water.
4. Incubate for 30 min in 12 mM $AgNO_3$.
5. Wash briefly with sterile water and then developing solution.
6. Incubate in fresh developing solution until RNA or protein bands of the desired intensity are observed.
7. Remove the developing solution, add 5% acetic acid, and incubate for at least 15 min to stop the reaction.

[a] The limit of detection for both RNA and protein is approximately 10 ng per 0.5 cm wide band in a 0.5 mm thick gel. Note that several snRNP proteins, including the U1-specific 70 kDa protein and the U5-specific 100 kDa protein, are visualized poorly by silver staining.

5. Reconstitution of snRNPs from purified RNA and protein

5.1 Introduction

Significant information regarding snRNP assembly and snRNA–protein interactions has been obtained from snRNP reconstitution studies performed in cell-free extracts (12–16). To facilitate further the structural, as well as functional, analysis of the spliceosomal snRNPs, the authors have recently developed an *in vitro* reconstitution system using purified snRNAs and native snRNP proteins (17). U1 and U2 snRNP complexes containing all the common snRNP proteins and various subsets of the particle-specific proteins have been reconstituted quantitatively. These particles exhibit physical properties (for instance, their buoyant densities in CsCl), which are similar to native particles, and they have been shown to be active substrates for cytoplasmic–nuclear transport in oocytes (18). One major advantage of this extract-free approach is that relatively homogeneous snRNP complexes can be obtained containing uniformly radiolabelled RNA. This greatly simplifies the analysis of protein–RNA interactions, for instance via UV cross-linking. Reconstitution in the absence of cell extracts and using purified components also simplifies studies on the assembly pathway and RNA–protein binding, since, in principle, reconstitution can be carried out with different subsets of snRNP proteins or with modified proteins generated by *in vitro* translation.

5.2 Preparation of snRNP protein

The disassembly method described for the signal recognition particle by Walter and Blobel (19) has been adapted for the isolation of snRNP proteins under non-denaturing conditions. The protein–RNA interactions within the snRNP particle are first weakened by chelation of divalent cations with EDTA and the proteins are subsequently separated from the RNA by ion-exchange chromatography on the polycationic resin, DE53. Optimal recovery of snRNA-free, snRNP proteins occurs when the disassembly is performed in buffer containing 0.15 M potassium acetate, 0.14 M NaCl, and 5 mM EDTA. A monovalent cation concentration less than 0.3 M is required to prevent release of snRNA from the DE53 resin.

Protocol 9 describes the preparation of native snRNP proteins. Under the conditions described in *Protocol 9*, all of the common snRNP polypeptides (B, B′, D1, D2, D3, E, F, G) and several of the particle-specific proteins are consistently recovered with an efficiency of 20–30%. However, the high molecular weight U5 proteins and the U1-specific 70 kDa protein are often lost. Subsequently, immunoaffinity chromatography (*Protocols 3* or *4*) can be used to remove selectively one or more of the individual proteins from the snRNP protein mixture. For efficient reconstitution of snRNPs using the purified proteins (Section 5.3), the RNA-free protein preparation must first be concentrated by dialysis against a 30% (w/v) polyethylene glycol (PEG) buffer. This method allows a 50–100-fold concentration to be consistently achieved without significant protein loss. *Protocol 9* has been used for the preparation of snRNP proteins from both individual snRNP species (such as those purified by Mono Q column chromatography; see *Protocol 5*) or from a mixture of snRNP particles (such as those isolated by anti-m_3G immunoaffinity chromatography; see *Protocol 3*).

Protocol 9. Preparation of native snRNP proteins

Equipment and reagents

- snRNP particles prepared as described in Protocols 3–5
- DE53 ion-exchange resin (Whatman)
- 4.0 M potassium acetate, pH 5.5
- Sterile water
- 4.0 M potassium acetate, pH 5.5
- 4.0 M NaCl
- 0.5 M DTE
- 0.5 M PMS
- Polyethylene glycol (PEG 6000; Merck)
- Sorvall centrifuge, HB4 rotor, and tubes (or equivalent)
- Wash buffer (0.15 M potassium acetate, pH 5.5, 0.14 M NaCl, 5 mM EDTA, 0.5 mM DTE, 0.5 mM PMSF); DTE and PMSF should be added directly before use.
- Reconstitution buffer (20 mM Hepes–KOH, pH 7.9, 50 mM KCl, 5 mM $MgCl_2$, 0.2 mM EDTA, 5% glycerol, 0.5 mM DTE, 0.5 mM PMSF); DTE and PMSF should be added directly before use.
- Reconstitution buffer containing 30% PEG 6000
- Dialysis tubing (3500 M_r exclusion)

Method

1. Prepare the DE53 resin by resuspending it in 10 vol. of 4.0 M potassium acetate, pH 5.5. Leave the resin for 5 min.
2. Wash the resin four times with 10 vol. of sterile water and then four times with 10 vol. of wash buffer.
3. Dilute the snRNP preparation to 133 μg/ml with 4 M potassium acetate, pH 5.5, 4.0 M NaCl, 0.5 M EDTA, 0.5 M DTE, 0.5 M PMSF, and sterile water such that the mixture ultimately contains 0.15 M potassium acetate, 0.14 M NaCl, 5 mM EDTA, 0.5 mM DTE, 0.5 mM PMSF[a].
4. Add the diluted snRNP preparation to the packed DE53 resin (2.5 ml packed resin per milligram snRNP) in a stoppered tube.
5. Incubate the mixture for 15 min on ice followed by 15 min at 37 °C, keeping the resin in suspension by inverting the tube once per min.
6. Centrifuge the mixture at 16 000 g for 10 min at 4 °C in the Sorvall HB4 rotor. Remove the supernatant and store it on ice.
7. Resuspend the DE53 resin in 1 vol. of wash buffer and incubate it for 15 min at 37 °C, mixing by inverting once per min as in step 5.
8. Centrifuge the resin and remove the supernatant as in step 6.
9. Combine the two supernatants (steps 6 and 8) and dialyse the pooled supernatant for 2 h at 4 °C against 50 vol. of reconstitution buffer.

Continued

Protocol 9. Continued

10. Dialyse the preparation at 4 °C against 30 vol. of reconstitution buffer containing 30% PEG 6000 until a 50–100-fold reduction in volume is achieved (typically 4–6 h)[b].
11. Dialyse the concentrated preparation for 15 min at 4 °C against reconstitution buffer.
12. Determine the protein concentration (ref. 7) and analyse both the protein and RNA by electrophoresis (*Protocols 6–8*).
13. Freeze the concentrated protein preparation in aliquots at −70 °C[c].

[a] The snRNP buffer constitution must be taken into consideration when calculating the final concentrations of the aforementioned buffer components.
[b] A final protein concentration greater than 0.2 µg/µl is required.
[c] An enhancement in the recovery of the U1-70 kDa protein and the U5 high molecular weight proteins has been observed if steps 9–11 are performed with a modified reconstitution buffer containing 20 mM KCl and 1.5 mM $MgCl_2$.

5.3 RNA preparation and snRNP reconstitution

Reconstitution is generally performed with individual snRNA species, isolated from native snRNP particles or generated by *in vitro* transcription (see Chapter 1, Section 3, for RNA isolation protocols). U1 snRNA transcribed *in vitro* has been used successfully to reconstitute core U1 snRNP particles, indicating that U1 snRNA post-transcriptional modifications are not essential for snRNP assembly. Optimal reconstitution of U1 and U2 snRNPs is observed when snRNA is incubated with a 5–10-fold molar excess of proteins over snRNA (for snRNP proteins, this corresponds to about 1 µg protein/pmol RNA). Reconstitution is typically performed in a buffer containing 5 mM $MgCl_2$ and 50 mM KCl, although little change in the efficiency of particle formation is observed if the former is between 2–15 mM and the latter between 50–250 mM. *Protocol 10* describes the generation of preparative amounts of reconstituted U1 core particles.

Protocol 10. Reconstitution of snRNP

Equipment and reagents

- snRNP protein purified by the method described in *Protocol 9*
- U1 snRNA (purified by denaturing gel electrophoresis as described in Chapter 1, *Protocol 8* or Chapter 4, *Protocol 11*)
- RNasin (Promega)
- 1.0 M Hepes-KOH, pH 7.9
- 2.0 M KCl
- 0.5 M $MgCl_2$
- 0.1 M DTE
- Beckman ultracentrifuge, TLA-100.3 rotor, and tubes (or equivalent)
- Buffered CsCl solution (20 mM Hepes-KOH, pH 7.9, 0.1 M KCl, 15 mM $MgCl_2$, 0.5 mM DTE, 0.5 mM PMSF, 0.7 g/ml CsCl)
- Refractometer

Continued

Protocol 10. *Continued*

Method

1. Mix 5 µg of U1 snRNA, 100 µg of purified snRNP protein, 1.0 M Hepes–KOH, pH 7.9, 2.0 M KCl, 0.5 M MgCl$_2$, 0.1 M DTE, and RNasin in a final volume of 0.2–0.5 ml, such that the final reconstitution mixture contains 20 mM Hepes–KOH, pH 7.9, 50 mM KCl, 5 mM MgCl$_2$, 1 mM DTE, and 0.5 units/µl of RNasin. Ensure that the protein concentration is at least 100 µg/ml.
2. Incubate the mixture for 30 min at 30 °C, followed by 15 min at 37 °C.
3. Layer the reconstitution mixture over 3 ml of buffered CsCl solution and centrifuge the gradient at 400 000 g for 16 h at 20 °C in a Beckman TLA-100.3 rotor.
4. Fractionate the gradient by manually removing 200 µl fractions from the top.
5. Analyse the RNA and protein content of each fraction by PAGE as described in *Protocols 6–8*. In this instance, 50 µl of each fraction is normally sufficient for analysis and should be diluted ~ 5-fold with sterile water before phenol extraction (see *Protocol 6*).
6. The buoyant density of the reconstituted particles may be determined by measuring the density of each fraction with a refractometer. The buoyant density of 12S U1 snRNPs is 1.4 g/cm^3.

For reconstitution at the analytical level, use radiolabelled snRNA, generated by 3' end-labelling (see Chapter 2, *Protocol 1* or Chapter 4, *Protocol 16*) or by *in vitro* transcription with a radiolabelled nucleotide (Chapter 1, *Protocol 4*). In this instance, reconstitution is essentially performed as described in *Protocol 10*, steps 1 and 2, except that the total reaction volume is 10–20 µl. Reconstitution with radiolabelled snRNA can be monitored by CsCl gradient centrifugation (see *Protocol 10*, steps 3 and 4). To detect the reconstituted particles and determine the efficiency of reconstitution, measure the amount of radioactivity in each fraction, as well as the density. Alternatively, the radiolabelled, reconstituted snRNPs may be analysed on native agarose gels (Chapter 1, *Protocol 6* or Chapter 4, *Protocol 12*). In the latter instance, reconstitution should be carried out in the presence of a 50-fold molar excess of carrier tRNA (add tRNA to the mixture in step 1 of *Protocol 10*).

6. Analysis of protein–RNA interactions by immunoprecipitation of snRNPs

6.1 Strategy

One of the most useful techniques for studying protein–nucleic acid interactions is immunoprecipitation. The association of a protein with a particular snRNA species is often initially confirmed by an immunoprecipitation assay using antibodies reacting specifically with one of the constituents of the corresponding RNP complex. In conjunction with RNA digestion methods (Sections 7 and 8), immunoprecipitation techniques are also used to map the binding site of a protein or group of proteins on the snRNA molecule. Immunoprecipitation utilizing Protein A coupled to Sepharose, first described by Kessler (20), was adapted by Lerner et al. (21) who, through the use of sera obtained from patients suffering from systemic lupus erythematosus, first demonstrated that snRNAs exist as RNP complexes with distinct protein components.

The preferred immunoprecipitation protocol involves initially adsorbing IgGs to PAS before the addition of the target antigen. This has several advantages over the addition of a crude antiserum to the RNP sample; most importantly, contaminating RNases can be removed by prewashing the PAS-bound antibody. After binding, the RNP particles are pelleted with the antibody–PAS complex and are washed extensively prior to further analysis.

Immunoprecipitation of RNP complexes may be performed with monoclonal or polyclonal antibodies directed against the protein of interest, provided the association of the antibody does not disrupt the protein–RNP interaction. Antibodies directed against other RNP constituents, such as modified nucleosides of the RNA or against stably-associated proteins (for instance, the proteins common to snRNPs), are also frequently used to monitor protein association. In both instances, a protein–RNP interaction, whether protein–RNA or protein–protein in nature, is established by the co-precipitation of the snRNA or stably-associated protein with the protein of interest. Analogous co-precipitation protocols, which make use of the well-characterized biotin–streptavidin interaction, are also often employed to monitor protein–RNA interactions. In this case, RNP complexes are precipitated with streptavidin–agarose or an equivalent matrix, typically by binding the snRNA molecule which has been biotinylated either directly (prior to RNP formation) or indirectly through the interaction of a biotinylated oligonucleotide (see Chapter 4, Section 3).

6.2 Immunoprecipitation of snRNPs

Protocol 11 describes a procedure for the precipitation of snRNPs using an antibody–PAS matrix.

Protocol 11. Immunoprecipitation of snRNPs

Equipment and reagents

- PAS[a] pre-swollen in PBS (2.5 mg of Protein A per 10 μl bed volume) prepared as described in *Protocol 1*.
- Antibody preparation (crude antiserum or hybridoma tissue culture supernatant)
- Non-immune serum or control antibody against a protein not associated with an snRNP
- PBS (20 mM potassium phosphate, pH 8.0, 0.13 M NaCl)

Continued

Protocol 11. *Continued*

- End-over-end rotary stirrer
- IPP$_{150}$ buffer (10 mM Tris–HCl, pH 8.0, 0.15 M NaCl, 0.1% Nonidet P-40)
- snRNP sample: generated by *in vitro* reconstitution and purified by either immunoaffinity chromatography (see *Protocols 3* and *4*) or ion-exchange chromatography on Mono Q (see *Protocol 5*), alternatively, a crude nuclear extract (Chapter 3, *Protocol 6*) may be used

Method

NB *Perform all steps at 4 °C*

1. Wash the preswollen PAS twice with 3 vol. of PBS, centrifuging at 10 000 g for 10 sec to pellet the PAS between the washes. Ten microlitres (bed volume) of PAS is required for each snRNP sample to be analysed.

2. Divide the washed PAS into two parts and pellet the PAS by centrifugation (10 000 g for 10 sec).

3. To one sample of PAS, add 5–20 µl of crude antiserum or 100–400 µl of hybridoma culture supernatant per 10 µl of pelleted PAS[b]. To the other PAS sample, add an equivalent amount of non-immune serum or a control antibody.

4. Mix each sample by end-over-end rotation overnight.

5. Pellet the antibody-bound PAS complexes, remove the supernatants and wash each of the PAS preparations four times with 5 vol. of IPP$_{150}$ buffer each time[c].

6. Resuspend each PAS in 5 vol. of fresh IPP$_{150}$ buffer and distribute aliquots of the suspensions, equivalent to 10 µl (bed volume), to individual 1.5 ml microcentrifuge tubes.

7. Pellet the PAS and discard the supernatant. Add the RNP sample. For quantitative immunoprecipitation, an excess of PAS-bound antibody is required. The maximum amount of a sample that may be adsorbed by 10 µl of PAS must be determined in a pilot experiment. As a guide, add a maximum of 5 µg of snRNP purified on Mono Q or 20 µl of nuclear extract (equivalent to 7×10^6 cells) per 10 µl of PAS containing anti-m$_3$G antibody or an antibody against a tightly-associated snRNP protein.

8. Distinguish true immunoprecipitation from background by setting up a parallel series of control incubations with the PAS linked to the non-immune serum or the control antibody. Adjust the sample volumes to a minimum of 0.3 ml by the addition of IPP$_{150}$ buffer. Incubate the samples for 2–4 h with rotary stirring.

9. Repeat step 5[d].

10. Resuspend each PAS sample in 5 vol. of IPP$_{150}$ buffer, transfer each to a new microcentrifuge tube, and pellet the PAS. Remove and discard the supernatant.

Continued

5: *Analysis of RNP interactions*

Protocol 11. *Continued*

11. Analyse the immunoprecipitated snRNP by one of the methods described in Section 6.3.

^a Protein G-Sepharose may be used with antibodies possessing a poor affinity for PAS.
^b When using IgM antibodies, such as monoclonal H386 against U1 snRNP 70 kDa protein, an anti-IgM antibody must first be bound to the matrix (20 μg/10 μl of PAS) as described in *Protocol 2* but without covalent coupling.
^c To reduce high background levels of RNP binding, higher stringency conditions can be used in steps 5–11, provided the protein of interest remains stably-associated and the antigen–antibody interaction is not disrupted. The most convenient way of increasing the stringency is to use IPP buffer containing 0.2–0.5 M NaCl.
^d When using IgM-linked PAS, it is recommended that the PAS is transferred to a new microcentrifuge tube after each wash to avoid high background levels. This is particularly important when washing at low stringency (i.e. <0.2 M NaCl).

6.3 Analysis of immunoprecipitated snRNPs

Immunoprecipitates from *Protocol 11* can be analysed in a number of ways depending on the experimental design. If immunoprecipitation has been performed with complexes containing a radiolabelled snRNA or protein (for instance, reconstituted U1 particles containing radiolabelled U1 RNA), the association of a given protein can be measured simply by determining the amount of radioactivity immunoprecipitated. This type of assay is referred to as a radioimmunoprecipitation assay. Alternatively, the immunoprecipitated protein and/or RNA can be isolated from the immunoprecipitate and analysed by PAGE. Proteins can be extracted by heating the PAS–antigen complex obtained from *Protocol 11* (step 11) in protein sample buffer at 90 °C for 10 min and analysed directly by SDS–PAGE as described in *Protocol 7*. A Western blot (22, 23) may be performed subsequently if the immunoprecipitated proteins cannot be detected by standard staining techniques.

To isolate immunoprecipitated snRNA, phenol extract the PAS complex, ethanol precipitate the RNA, and analyse it by denaturing polyacrylamide gel electrophoresis as described in *Protocol 6* and Chapter 1, *Protocol 8* or Chapter 4, *Protocol 11*. The fractionated RNA may be detected directly by staining or by autoradiography if it has been radiolabelled prior to electrophoretic analysis, for instance by 3' end-labelling with [^{32}P]pCp (see Chapter 4, *Protocol 16* or Volume II, Chapter 3, *Protocol 4*).

When the antibody–antigen complex can be disrupted by the addition of an excess of competing peptide or nucleoside, immunoprecipitated material can be eluted specifically and subsequently analysed (as above); in this way, background levels can also be reduced significantly. The results of an immunoprecipitation assay, in which the interaction of snRNP proteins during reconstitution of U1 and U2 snRNPs was monitored by analysing the precipitated radioactive snRNAs on denaturing polyacrylamide gels, are presented in *Figure 3*.

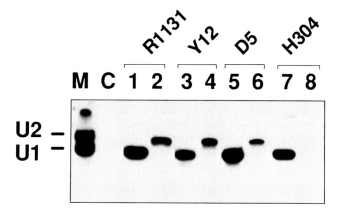

Figure 3. Immunoprecipitation of reconstituted U1 and U2 snRNPs. U1 and U2 RNP particles were reconstituted with ^{32}P-end-labelled U1 RNA (odd numbered lanes) or ^{32}P-end-labelled U2 snRNA (even numbered lanes) essentially as described in *Protocol 10*. Immunoprecipitation was performed as described in *Protocol 11* with the following antibodies directed against snRNP-specific components: R1131 (anti-m$_3$G), Y12 (anti-Sm, i.e. a subset of the common snRNP proteins), D5 (anti-A and B" proteins), and H304 (anti-A protein). The immunoprecipitated RNA was isolated and fractionated on a 10% polyacrylamide/7 M urea gel and visualized by autoradiography (see Chapter 1, *Protocol 8*). Lane M, purified U1 and U2 RNA; lane C, background (negative) control (immunoprecipitation carried out with an antibody directed against a ribosomal protein). The authors thank Utz Fischer for providing this figure.

7. Analysis of RNA–protein interactions by UV cross-linking

7.1 Strategy

Cross-linking analyses have proved highly useful for the elucidation of RNP higher order structure. For example, the pioneering cross-linking analyses of ribosomal subunits by Brimacombe *et al.* (24) have contributed significantly to the elucidation of intermolecular interactions within these RNP complexes. RNA–RNA, RNA–protein, and protein–protein interactions can be identified by their stabilization through the formation of covalent cross-links. Cross-links can be induced either by bifunctional chemical reagents or by direct irradiation with ultraviolet light. Cross-linking with UV light has several advantages over chemical reagents; it is relatively easy to carry out and the dose level can be accurately controlled. Irradiation at 254 nm, for example, allows the identification of direct RNA–protein interactions since cross-link formation is observed only between molecular entities which are not separated by more than one covalent bond length (25). Due to the differential absorption of RNA and protein at this wavelength, sufficient RNA–protein cross-links can be induced whilst avoiding a significant level of intra-protein cross-links or seriously degrading the RNA and protein (26). After UV irradiation, several techniques (including RNase digestion, two-dimensional PAGE, immunoprecipitation, and RNA fingerprinting) may be carried out to identify the precise protein and RNA sequences which have been covalently linked.

7.2 snRNP preparation and UV irradiation

Detailed cross-link analyses require snRNP particles containing uniformly radiolabelled RNA. Radiolabelled snRNP particles can be obtained by labelling cells *in vivo* by the addition of [^{32}P]*ortho*phosphate to the tissue culture medium. Alternatively, radiolabelled snRNAs can be generated by transcription *in vitro* in the presence of ^{32}P-labelled ribonucleoside triphosphates and then used to form radiolabelled snRNP particles by *in vitro* reconstitution (see *Protocol 10*). The latter method is recommended since it is difficult through labelling *in vivo* to generate snRNA with a specific radioactivity which is sufficiently high for RNA fingerprint analysis.

There is also another potential disadvantage to using labelled particles which have been obtained through labelling *in vivo*. Upon irradiation, snRNP proteins directly bound to the RNA are covalently attached to one or more radioactively labelled nucleotides and, thereby, they too become radioactively labelled. If any of the snRNP proteins have been labelled already *in vivo* by phosphorylation, their presence will complicate the initial interpretations regarding the identity of the cross-linked protein(s). That is, it has to be assumed that after irradiation, the only radioactive species observed, aside from RNA alone, are proteins cross-linked to the RNA. Thus, if snRNP labelled *in vivo* is to be used, it is essential to ascertain in advance whether radiolabelled snRNP proteins are already present due to protein phosphorylation.

Protocol 12 describes the conditions required for the generation of cross-links in snRNP particles prepared *in vitro*. This procedure has also been performed successfully with purified particles labelled *in vivo*. The UV dose range recommended in *Protocol 12* is 10–40 mJ/mm^2. If different conditions or another type of RNP particle are used, construct a dose-titration curve to establish the UV dose yielding a sufficient number of RNA–protein cross-links with a minimal amount of protein or RNA damage. The following controls are included in *Protocol 12*: non-irradiated snRNP particles, and both irradiated and non-irradiated radiolabelled RNA. In addition, it may be useful, for comparative purposes, to include radiolabelled 'mutant' snRNA which lacks a suspected region of interaction.

Protocol 12. UV irradiation of snRNP particles

Equipment and reagents

- RNP particles reconstituted as described in *Protocol 10*, using snRNA transcribed *in vitro* to a specific activity of 1–3 × 10^6 c.p.m./pmol (see Chapter 1, *Protocol 4* or Chapter 4, *Protocol 14*). For control cross-linking, a sample of the radiolabelled RNA is also required.

Continued

Protocol 12. *Continued*

- Microtitre plates (Greiner; ELISA plates, F-form)
- Sylvania G8T5 germicidal UV lamp (or equivalent)

Method

NB *Wear safety goggles to protect the eyes when working with UV light*

1. Dispense aliquots of the snRNP particle preparation (20–50 µl) into the wells of two microtitre plates to give a solution depth of about 1 mm.
2. In other wells of the microtitre plates, dispense aliquots of the radiolabelled snRNA equivalent to the amount of radioactivity present in the samples of snRNP particles.
3. Place *one* of the microtitre dishes at a distance of 4 cm from the UV source.
4. Irradiate the plate (without its cover) at 254 nm for 2–4 min[a]. Do not irradiate the other (control) plate.
5. Transfer the samples from *both* microtitre plates to separate microcentrifuge tubes. Analyse the cross-linked snRNP particles immediately as described in Sections 7.3–7.5 or store them at −70 °C until assayed.

[a] Allow the UV source to warm up for at least 30 min prior to use to ensure a constant UV dose.

7.3 Analysis of the extent of UV cross-linking

To determine the extent of cross-linking, subject the samples from *Protocol 12* (step 5) to either denaturing polyacrylamide–urea gel electrophoresis (Chapter 1, *Protocol 8* or Chapter 4, *Protocol 11*) or SDS–PAGE (see *Protocol 7*). The presence of radioactive species migrating more slowly than the controls (i.e. the non-irradiated snRNP particles and both irradiated and non-irradiated snRNA) is an indication of cross-link formation. In addition, a distinction can be made between intra-RNA and protein–RNA cross-links by comparing cross-linked particles with and without protease digestion. Perform the protease digestion by incubating the irradiated samples from *Protocol 12* (step 5) with 10 µg of proteinase K (Merck) for 30 min at 37 °C in the presence of 0.5% (final concentration) of SDS. The identities of the proteins cross-linked to the snRNA and the precise sites of attachment to the RNA may be established by the methods described in the following sections (Sections 7.4 and 7.5).

7.4 Identification of cross-linked proteins

To facilitate the identification of the cross-linked polypeptide, the large RNA–protein complexes generated by cross-linking are reduced to protein–oligonucleotide complexes by digestion with a mixture of RNases of differing specificity. Multiple cross-link sites are thereby separated physically and the size of the RNA

5: Analysis of RNP interactions

oligonucleotide attached to the protein is reduced to 1–3 nt. In this way, the cross-linked protein exhibits only slightly altered electrophoretic mobility when subsequently subjected to SDS–PAGE or two-dimensional gel electrophoresis and its probable identity can be deduced more readily. *Protocol 13* describes the digestion of cross-linked snRNP particles using a mixture of nucleases A, T1, S7, and T2 essentially as specified by Stiege *et al.* (27).

Protocol 13. Digestion of cross-linked snRNP particles using multiple RNases

Reagents

- Cross-linked snRNP particles (see *Protocol 12*)
- 0.5 mM DTE, 0.5 mM PMSF, 2 mg/ml leupeptin (Boehringer Mannheim) in sterile water
- RNase A (Boehringer Mannheim)
- RNase T1 (PL Biochemicals)
- RNase T2 (Calbiochem)
- Nuclease S7 (micrococcal nuclease; Boehringer Mannheim)
- Nonidet P-40 (NP-40)
- 1.0 M $CaCl_2$
- Urea
- 0.2 M sodium acetate buffer, pH 4.0

Method

1. Dilute 10 μg of snRNPs from *Protocol 12*, step 5, (in 25 μl reconstitution in mixture; *Protocol 10*, step 1) to 100 μl with sterile water containing 0.5 mM DTE, 0.5 mM PMSF and 2.0 μg/ml leupeptin.
2. Add 1 μg RNase A, 7 units of RNase T1, 75 units of nuclease S7, NP-40 to 0.5% (v/v), and 1.0 M $CaCl_2$ to a final concentration of 10 mM.
3. Incubate the mixture at 37 °C for 45 min.
4. Add solid urea to a final concentration of 4.0 M.
5. After the urea has dissolved, incubate the mixture at 37 °C for 20 min.
6. Increase the urea concentration to 7.0 M and incubate the mixture at 37 °C for an additional 20 min.
7. Increase the temperature gradually over a 45 min period from 37 °C to 60 °C.
8. Incubate at 60 °C for 5 min, then cool the samples to 37 °C and adjust the pH to 4.5 by adding 30 μl of 0.2 M sodium acetate, pH 4.0.
9. Add 2.5 units of RNase T2 and incubate at 37 °C for 45 min.
10. Repeat step 7.
11. Add 5 vol. of 95% ethanol.
12. Allow the protein–oligonucleotide complexes to precipitate by incubating the samples at −70 °C for 1 h or at −20 °C overnight.
13. Centrifuge the samples at 12 000 g for 10 min, wash the precipitates with 80% (v/v) ethanol and vacuum desiccate them.
14. Analyse the cross-linked samples by SDS–PAGE as described in *Protocol 7*, or store them at −70 °C until assayed.

Significant information regarding the identity of a cross-linked protein can be obtained through SDS–PAGE by comparing the M_r of the protein–oligonucleotide with the known M_r of the snRNP proteins. While every protein whose M_r is lower than the protein–oligonucleotide complex is a potential cross-link candidate, a protein whose M_r is higher than that of the unidentified protein–oligonucleotide clearly cannot be the source of the cross-linked protein. To further limit the number of potential candidates, two-dimensional PAGE can be performed. In this case comparisons are made of both the apparent pI and the M_r of the radiolabelled protein–oligonucleotide complex with those of normal, uncomplexed snRNP protein. Since the covalent linkage of an oligonucleotide to a polypeptide will change the polypeptide's pI to a more acidic value, only those proteins which normally possess pI values more basic than the protein–oligonucleotide complex are potential cross-link candidates; those with more acid pIs can be discounted. The precise identity of the cross-linked protein may be established ultimately by radioimmunoprecipitation of the oligonucleotide–protein complex with antibodies directed against the candidate protein (Section 6) or by Western blotting (22, 23).

7.5 Identification of RNA cross-link sites

Rapid localization of the precise site of cross-link formation in an RNA molecule is readily achieved by RNA fingerprint analysis. A two-dimensional array of nuclease-resistant oligonucleotides (the fingerprint) is generated by subjecting the RNA to nuclease digestion and then separating the resultant oligonucleotide fragments by two-dimensional thin-layer chromatography. Oligonucleotides containing covalently-attached peptides exhibit a significantly different migration behaviour from a free oligonucleotide of identical sequence. Thus, those oligonucleotide fragments containing cross-linked amino acids can be identified by comparing the pattern obtained from snRNP samples subjected to UV irradiation with that of a non-irradiated control sample (see *Protocol 12*, step 4). Detailed protocols for RNA fingerprinting will be found in another volume of this Series (28). Prior to the fingerprint analysis, the cross-linked complexes must first be purified by gel electrophoresis to remove free RNA and then subjected to protease digestion to generate RNA–oligopeptides. *Protocol 14* describes this preliminary treatment of the sample before RNA fingerprinting is carried out.

Protocol 14. Purification and protease digestion of cross-linked snRNP complexes

Equipment and reagents

- Reconstituted snRNP preparation under investigation, cross-linked by UV irradiation (see *Protocol 12*)
- 10% polyacrylamide/7.0 M urea/0.1%
- SDS denaturing gel, dyes-loading buffer, TBE buffer, and equipment as specified in Chapter 1, *Protocol 8* or Chapter 4, *Protocol 11*

Continued

Protocol 14. *Continued*

- Extraction buffer (80 mM ammonium carbonate, 0.1% SDS)
- 3.0 M sodium acetate, pH 5.5
- 10 mM Tris–HCl, pH 7.8, 0.1% SDS, 1 mM EDTA
- 1.0 M NaCl
- Proteinase K (Merck)
- 10 mg/ml carrier tRNA and phenol: chloroform:isoamylalcohol mixture (see *Protocol 6*)

Method

1. Load the UV-irradiated sample on to a polyacrylamide/7 M urea/0.1% SDS gel and fractionate the sample as described in Chapter 1, *Protocol 8* or Chapter 4, *Protocol 11*.
2. Locate the slower-migrating cross-linked complexes by autoradiography. Cut out the corresponding gel fragment.
3. Elute the cross-linked complexes by incubating the gel fragment overnight at 4 °C in 10 vol. of extraction buffer.
4. Add 10 vol. of acetone.
5. Process the samples as described in *Protocol 13*, (steps 12 and 13).
6. Redissolve the sample in 0.2 ml of 10 mM Tris–HCl, pH 7.8, 0.1% SDS, 1 mM EDTA.
7. Add an equal vol. of 1 M NaCl and Proteinase K to a final concentration of 1 mg/ml.
8. Incubate the mixture for 30 min at 37 °C.
9. Add 0.1 vol. of carrier tRNA (10 mg/ml) and extract the mixture with an equal vol of phenol:chloroform:isoamylalcohol mixture (see *Protocol 6*, steps 1 and 2).
10. Remove the upper (aqueous) phase, add 2 vol. of absolute ethanol and process the sample as described in *Protocol 13* (steps 12 and 13).

8. Mapping of RNA–protein interactions by RNase protection

The higher order structure of an RNP particle can be probed with nucleases or chemical reagents which modify RNA. Those RNA regions susceptible to nuclease digestion, for example, are assumed to be free of protein or not involved in an interaction with another RNP particle. The regions of RNA which are protected from RNase digestion can be isolated subsequently (for instance, by gel electrophoresis and elution) and the sequence of the protected region determined by RNA fingerprint analysis. For example, micrococcal nuclease protection experiments in conjunction with RNA fingerprinting were used to map the RNA binding site of the common snRNP proteins (the so-called Sm-site or domain A) (29). The combination of RNase digestion followed by immunoprecipitation allows the characterization of proteins interacting with the protected RNA fragment. To facilitate this detection and to permit subsequent RNA fingerprint analysis, RNP particles containing radiolabelled RNA (generated as described in Section 7.2) are normally employed. In a standard RNase protection experiment, deproteinized snRNA is included as a positive control for digestion. Protection patterns from various snRNP sources are often compared in order to detect changes in the protein composition of a particular particle or in its association with other RNP particles (for instance, with the spliceosome). RNases of restricted specificity (such as RNase T1), which typically generate larger RNA–protein fragments, may also be employed.

Protocol 15 describes the digestion of snRNPs with micrococcal nuclease. The protocol is adapted for Liautard *et al.* (29) and has been used successfully to monitor the interaction of the common snRNP proteins at the U1 Sm-site during snRNP reconstitution. The concentration of micrococcal nuclease suggested in *Protocol 15* is generally satisfactory but should be checked in a pilot experiment.

Protocol 15. Micrococcal nuclease digestion of snRNPs

Equipment and reagents

- Reconstituted snRNP preparation (in reconstitution buffer at an RNA concentration of ~50 µg/ml) and deproteinized snRNA used for the reconstitution of snRNP (see *Protocol 10*)
- Micrococcal nuclease (Boehringer Mannheim; 20 000 units/ml)
- 0.1 M $CaCl_2$
- 50 mM EGTA
- 10 mg/ml carrier tRNA and phenol:chloroform:isoamylalcohol mixture as described in *Protocol 6*.
- 20% polyacrylamide/7.0 M urea denaturing gel, dyes-loading buffer, TBE buffer, and electrophoresis equipment (Chapter 1, *Protocol 8* or Chapter 4, *Protocol 11*)

Method

1. Add 0.1 M $CaCl_2$ and micrococcal nuclease (final concentration of 1 mM and 5000 units/ml, respectively) to the snRNP and snRNA samples[a].
2. Incubate the samples at 37 °C for 30 min.
3. Terminate the reactions by adding 50 mM EGTA to a final concentration of 2 mM.
4. Add carrier 10 mg/ml tRNA to 10 µg/ml and phenol extract the samples as described in *Protocol 6*, steps 1 and 2.
5. Remove the upper (aqueous) phase and recover the RNA from the samples as described in *Protocol 6*, steps 4, 5, and 8–10.
6. Analyse the protected RNA fragments by fractionating the RNA samples on a 20% polyacrylamide/7 M urea gel as described in Chapter 1 *Protocol 8* or Chapter 4, *Protocol 11*.

[a] Samples containing radiolabelled snRNA should be used.

5: *Analysis of RNP interactions*

9. Immunoelectron microscopy of snRNP particles

9.1 Introduction

The pioneering studies on ribosomal subunit structure by Lake and Stöffler and their co-workers (30, 31) demonstrated the power of electron microscopy for the determination of the three-dimensional structure of RNP particles, as well as the spatial arrangement of the individual components. The snRNPs with sedimentation coefficients between 10S and 20S, are structurally similar to ribosomal subunits and thus can be readily investigated by classical electron microscopy and immunoelectron microscopy. Individual snRNP particles have been shown by electron microscopy to be structurally asymmetric (32–34) and, therefore, lend themselves to immunoelectron microscopy in which IgG antibodies are used to mark the position of individual snRNP components. In this way, it is possible to localize the relative positions of individual snRNP proteins, snRNA sequences, cofactors, and even the interaction sites of a particular snRNP with pre-mRNA or other snRNPs within the spliceosome.

Since electron microscopy is a highly specialized procedure normally performed in collaboration with skilled electron microscopists, experimental protocols will not be provided here. Rather this section concentrates on sample preparation and provides advice for ensuring successfully electron microscopy.

9.2 Stabilization of snRNP–protein interactions for immunoelectron microscopy

The association of a number of snRNP proteins with their cognate RNP particle is not sufficiently stable for immunoelectron microscopy. To prevent their dissociation during the isolation of the immunocomplex and subsequent preparation of the sample for electron microscopy, the snRNP particles may be subjected to chemical cross-linking prior to formation of the immunocomplex. Dithio-bissuccinimidyl propionate (DSP), which cross-links primary and secondary amino groups, has been used successfully to covalently cross-link U1 snRNP proteins, without significantly altering the shape of the particle (35). The cross-linking procedure described in *Protocol 16* is adapted from Lomant and Fairbanks (36).

Protocol 16. DSP cross-linking of snRNPs

Reagents

- snRNP particles (>150 μg protein/ml) prepared as described in *Protocols 3–5*
- Cross-linking buffer (20 mM triethanolamine–HCl, pH 8.5, 0.3 M KCl, 1.5 mM $MgCl_2$)
- 18 mM DSP reagent in DMSO (Pierce; prepare directly before use)
- 3.0 M glycinamide–HCl, pH 4.0 (Sigma)
- Reagents for SDS–PAGE of proteins (see *Protocol 7*)

Continued

Protocol 16. *Continued*

Method

1. Dialyse the snRNP preparation for 4 h at 4 °C against 50 vol. of cross-linking buffer[a].
2. Check the protein concentration of the snRNP sample and dilute it, if necessary, to 150–300 µg/ml with cross-linking buffer.
3. Slowly add 18 mM DSP until the final concentration of DSP is 60 µM.
4. Incubate the mixture at 0 °C for 30 min.
5. Stop the reaction by adding 3 M glycinamide–HCl, pH 4.0, to 50 mM.
6. Incubate the mixture at 37 °C for 40 min.
7. Monitor the extent of cross-link formation by SDS–PAGE. Load a volume of the snRNP particles equivalent to 10 µg of proteins as described in *Protocol 7* (except sample buffer without DTE must be used). Cross-linking is evidenced by a decreased mobility of the snRNP proteins.

[a] Alternatively, if the U1 snRNPs are initially purified by glycerol gradient centrifugation (see *Protocol 4*), prepare the gradients with cross-linking buffer and omit step 1 of this protocol.

The cross-linked snRNP particles from *Protocol 16* can be used directly in the formation of immuno–snRNP complexes (see *Protocol 17*) and/or electron microscopy.

9.3 Formation of antibody–snRNP complexes

Antibodies of the IgG class are used exclusively for immunoelectron microscopy since their characteristic Y-shape can generally be easily distinguished. Although both polyclonal and monoclonal antibodies have been successfully used for investigating the higher order structure of ribosomal subunits, monoclonal or affinity-purified polyclonal antibodies are recommended for immunoelectron microscopy. Immunocomplexes are generated by incubating purified snRNPs, such as those purified by Mono Q chromatography or immunoaffinity methods (see *Protocols 3–5*), with an equimolar amount of specific antibody. The precise amount of a particular antibody required for efficient formation of complexes must be determined in pilot experiments. To control for non-specific antibody interactions, it is also advisable to analyse control complexes formed either with snRNP particles which lack the target antigen or in the presence of an excess of the purified target antigen (such as a peptide or modified nucleoside). Although electron microscopy preparations can be made directly from the antibody–snRNP reaction mixture, identification of the immunocomplexes in electron micrographs is often greatly facilitated if any significant amount of the free IgG present is first removed by centrifugation on glycerol or sucrose gradients. *Protocol 17* describes conditions for the generation of complexes between U1 snRNP and antibodies against the U1-specific A protein.

5: Analysis of RNP interactions

Protocol 17. Formation of snRNP–antibody complexes

Equipment and reagents

- U1 snRNPs cross-linked using DSP as described in *Protocol 16*
- D5 monoclonal antibody (IgG) against U1 A protein
- S-buffer (20 mM Hepes–KOH, pH 7.9, 0.15 M KCl, 1.5 mM $MgCl_2$, 0.5 mM PMSF). Add PMSF directly before use
- 40% sucrose
- 6% and 20% sucrose solutions made in S-buffer using the 40% sucrose stock solution
- Beckman ultracentrifuge, TLS-55 rotor and thick-walled polycarbonate tubes (or equivalent)
- Gradient former and fractionator
- PBS (20 mM potassium phosphate, pH 8.0, 0.13 M NaCl)

Method

1. Combine 10 μg of U1 snRNPs cross-linked using DSP with 4.3 μg of the A protein-specific monoclonal IgG antibody (D5) in a maximum of 25 μl of PBS.

2. Incubate the mixture overnight[a] at 4 °C.

3. Dilute the sample to 0.1 ml with S-buffer.

4. Prepare 1.4 ml linear 6–20% sucrose gradients in S-buffer using thick-walled polycarbonate tubes for the TLS-55 rotor (or equivalent)[b].

5. Load the diluted snRNP–antibody mixture on to a 1.4 ml sucrose gradient and centrifuge the gradient at 4 °C in the Beckman TLS-55 rotor at 166 000 g for 6 h[c].

6. Fractionate the gradient into 0.12 ml fractions.

7. Facilitate immunocomplex detection and quantitation by performing ELISA on each fraction to determine their snRNP and antibody content (see *Protocol 18*).

8. Process the snRNP–IgG complexes from the peak snRNP–IgG fraction for electron microscopy as described in Section 9.3. If the complexes are not directly processed, freeze them in liquid nitrogen and store them at −70 °C.

[a] The optimal incubation time for a given antibody–snRNP interaction must be determined in pilot experiments.
[b] The volume of the gradient must be limited to prevent significant dilution of the snRNP samples and this is why thick-walled, polycarbonate tubes are used.
[c] If larger snRNP particles (e.g. 20S U5 snRNPs) are being studied, separate them from IgGs on 10–30% glycerol gradients prepared in S-buffer. Centrifuge these gradients at 166 000 g for 4.5 h.

Protocol 18. ELISA of gradient fractions

Equipment and reagents

- Microtitre plates (Greiner, ELISA F-form)
- Automated ELISA reader
- BSA (Sigma, RIA grade)
- PBS (20 mM potassium phosphate, pH 7.4, 0.13 M NaCl)
- ELISA buffer (0.1% Tween-20 in PBS buffer)
- ELISA buffer containing 1.0% BSA
- 1.0% BSA prepared in PBS
- Human anti-RNP serum
- Phosphatase-conjugated anti-human IgG antibody (Sigma). Dilute this with ELISA buffer containing 1.0% BSA directly before use
- Phosphatase-conjugated anti-mouse IgG antibody (Sigma)
- Substrate buffer (1 mg/ml *para*-nitrophenyl phosphate, 100 mM sodium carbonate, pH 9.5, 2 mM $MgCl_2$). Prepare this directly before use

Method

1. Pipette 40 µl of each sucrose gradient fraction (from *Protocol 17*, step 6) and 10 µl of PBS into two separate ELISA microtitre plates and incubate both plates at 4 °C overnight. Include a background control (50 µl of PBS). Label the plates A (snRNP) and B (IgG).
2. Wash the wells with PBS. The solutions are best removed from the ELISA plate by inverting and shaking, followed by blotting on a paper towel.
3. Add 100 µl of 1.0% BSA in PBS to each well and incubate the plates for 2 h at room temperature or overnight at 4 °C.
4. Wash both plates with PBS. Repeat twice.
5. Dilute the human anti-RNP serum 1:1000 in ELISA buffer containing 1.0% BSA. Pipette 80 µl of the anti-RNP solution into the wells of plate A and 80 µl of ELISA buffer containing 1.0% BSA in the wells of plate B. Incubate the plates for 2 h at room temperature or overnight at 4 °C.
6. Wash plates A and B with ELISA buffer. Repeat twice.
7. Dilute the phosphatase-conjugated anti-human and anti-mouse IgG antibodies 1:1000 in ELISA buffer.
8. Pipette 80 µl of the phosphatase-conjugated anti-human IgG antibody into the wells of plate A and 80 µl of the phosphatase-conjugated anti-mouse IgG antibody into the wells of plate B. Incubate the plates for 2 h at room temperature.
9. Wash both plates with ELISA buffer. Repeat twice.
10. Prepare substrate buffer and add 80 µl of it to the wells of the microtitre plates.
11. Incubate the plates at room temperature for 30–240 min. During this time measure the A_{405} and A_{450} with an ELISA reader.
12. Determine the relative phosphatase activity bound to the plate by subtracting the A_{450} from A_{405}. Plot the activity values over the fraction numbers and estimate the relative amounts of free IgG, snRNP, and IgG–snRNP complexes. Typically, the peak of free IgG is in fraction 5 (from the bottom) while free snRNP is in fraction 7. SnRNP–IgG complexes sediment slightly faster than the free particles.

9.4 Preparation of samples for electron microscopy

For visualizing snRNPs and snRNP–IgG complexes negative staining by the double carbon film method is recommended. In this procedure, snRNPs are first adsorbed on to a carbon film, contrasted with uranyl formate and then sandwiched with a second carbon film. Electron microscopic imaging of negatively-contrasted snRNPs allows clear visualization of the outline of the particles. In addition, this procedure is particularly well-suited for the visualization of bound IgG, since all regions of this Y-shaped molecule are forced into a coplanar orientation, thereby facilitating its identification. A detailed procedure for the preparation of negatively-contrasted ribosomes, which has been adapted for use with snRNPs, is described by Stöffler-Meilicke and Stöffler (37). This reference should be consulted for experimental details. The following points apply when preparing negatively-contrasted snRNPs:

(a) Carbon films best suited for snRNP adsorption are generated by indirect, as opposed to direct, deposition of carbon vapour on mica.
(b) The optimal concentration of snRNP is 15–30 µg protein/ml.
(c) The optimal adsorbtion time at 4 °C varies from 20 sec to 2 min and must be determined in pilot experiments.

9.5 Investigation of snRNPs and snRNP–IgG complexes by electron microscopy

Image the snRNPs with an acceleration voltage of 80 kV. Prepare the electron micrographs at a primary magnification of 65 000–80 000 for large snRNPs or 100 000–140 000 for small snRNPs. For subsequent examination, photographically enlarge the micrographs to a final magnification of 300 000–500 000.

The interpretation of immunocomplex images generated by electron microscopy is described in detail for ribosomes by Stöffler-Meilicke and Stöffler (37) and that for snRNPs is analogous. Note that due to the presence of two antigen binding sites per IgG molecule, both binary snRNP–IgG and ternary snRNP–IgG–snRNP complexes may be observed.

In *Figures 4* and *5*, examples of U1 snRNPs or U1 snRNP:anti-A protein antibody complexes are shown. Negatively-stained U1 snRNPs (see *Figure 4*) possess an almost circular main body, about 8 nm in diameter, with two characteristic protuberances, 4–7 nm long and 3–4 nm wide. The apparent variation in the observed form of the U1 particle arises mainly from different particle orientations on the carbon film; that is, they represent different two-dimensional projections of a three-dimensional object. The ability to recognize the IgG binding site on the U1 particle requires some practice since both the U1 snRNP and IgG possess a similar shape and size. The two components can be distinguished, however, since the U1 snRNP protuberances are somewhat

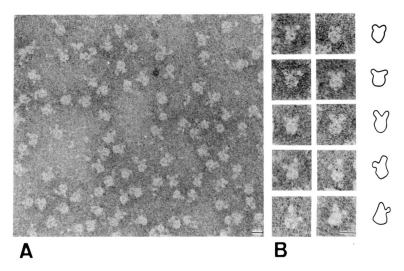

Figure 4. U1 snRNPs visualized by electron microscopy after negative staining with uranyl formate. A. Representative view. B. Gallery of particles, orientated such that the protuberances point upwards. The images in each row are interpreted in the drawing on the right. Bar, 10 nm.

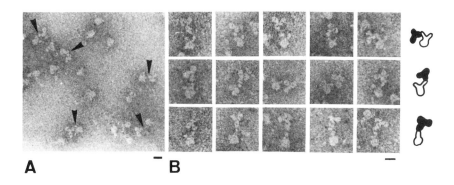

Figure 5. Binding of anti-A protein antibodies to U1 snRNPs visualized by electron microscopy. Immunocomplexes were formed by incubation of DSP-cross-linked U1 snRNPs with the A protein-specific monoclonal antibody, D5, separated on sucrose gradients (see *Protocol 17*) and visualized by electron microscopy (see Sections 9.4 and 9.5). A. General view of the sucrose gradient fraction containing the snRNP-IgG complexes (arrowheads point to the antibody-binding sites). B. Gallery of selected U1-(anti-A IgG) immunocomplexes, with the U1 protuberances oriented upright. The complex at the right in each row is interpreted in the adjacent sketch (the black area marks the antibody). Bar, 10 nm.

thinner than the antibody arms. As depicted in *Figure 5*, the binding site of the anti-A protein antibody and, hence the position of the A protein, can be localized to one of the U1 particle's protuberances. The relative position of this protein within the U1 snRNP can be further defined by double antibody labelling experiments which are described in detail by Kastner *et al.* (35).

Acknowledgements

Work from the authors' laboratory has been supported by grants from the Deutsche Forschungsgemeinschaft (SFB 272/A3 and SFB 286/A4), the Ministry of Science and Technology (BMFT, Marburg), and the Fonds der Deutschen Chemischen Industrie. Cindy L. Will was supported in part by a fellowship from the Alexander von Humboldt Stiftung. The authors thank the following members of their laboratory for helpful discussions and critical reading of the manuscript: Silk Börner, Wolfgang Hackl, Volker Heinrichs, Ute Kornstädt and Irene Öchsner-Welpelo.

References

1. Lührmann, R., Kastner, B., and Bach, M. (1990). *Biochim. Biophys. Acta Gene Struct. Expression*, **1087**, 265.
2. Hoch, S. O. (1990). In *Methods in enzymology*, (ed. J. E. Dahlberg and J. N. Abelson), Vol. 181, pp. 257–263. Academic Press, San Diego.
3. Harlow, E. and Lane, D. (ed.) (1988). *Antibodies, a laboratory manual.* Cold Spring Harbor Press, Cold Spring Harbor, NY.
4. Bach, M., Bringmann, P., and Lührmann, R. (1990). In *Methods in enzymology*, (ed. J. E. Dahlberg and J. N. Abelson), Vol. 181, pp. 232–257. Academic Press, San Diego.
5. Behrens, S.-E. and Lührmann, R. (1991). *Genes Dev.*, **5**, 1439.
6. Dignam, J. D., Lebovitz, R. M., and Roeder, R. G. (1983). *Nucleic Acids Res.*, **11**, 1475.
7. Bearden, J. C. (1978). *Biochim. Biophys. Acta*, **533**, 525.
8. Krämer, A. (1990). In *Methods in enzymology*, (ed. J. E. Dahlberg and J. N. Abelson), Vol. 181, pp. 215–232. Academic Press, San Diego.
9. Nelissen, R. L. H., Heinrichs, V., Habets, W. J., Simons, F., Lührmann, R., and van Venrooij, W. J. (1991). *Nucleic Acids Res.*, **19**, 449.
10. Lehmeier, T., Foulaki, K., and Lührmann, R. (1990). *Nucleic Acids Res.*, **18**, 6475.
11. Merril, C. R., Goldman, D., Sedman, S. A., and Ebert, M. H. (1981). *Science*, **211**, 1437.
12. Hamm, J., Kazmaier, M., and Mattaj, I. W. (1987). *EMBO J.*, **6**, 3479.
13. Hamm, J., Van Santen, V. L., Spritz, R. A., and Mattaj, I. W. (1988). *Mol. Cell. Biol.*, **8**, 4787.
14. Patton, J. R., Habets, W., van Venrooij, W. J., and Pederson, T. (1989). *Mol. Cell. Biol.*, **9**, 3360.
15. Patton J. R., Patterson, R. J., and Pederson, T. (1987). *Mol. Cell. Biol.*, **7**, 4030.
16. Pikielny, C. W., Bindereif, A., and Green, M. R. (1989). *Genes Dev.*, **3**, 479.

17. Sumpter, V., Kahrs, A., Fischer, U., Kornstädt, U., and Lührmann, R. (1992). *Mol. Biol. Rep.*, **16**, 229.
18. Fischer, U., Sumpter, V., Sekine, M., Satoh, T., and Lührmann, R. (1993). *EMBO J.*, **12**, 573.
19. Walter, P. and Blobel, G. (1983). *Cell*, **34**, 525.
20. Kessler, S. W. (1975). *J. Immunol.*, **115**, 1619.
21. Lerner, M. R. and Steitz, J. A. (1979). *Proc. Natl Acad. Sci. USA*, **76**, 5495.
22. Towbin, H., Staehlin, T., and Gordon, J. (1979). *Proc. Natl Acad. Sci. USA*, **76**, 4350.
23. Zeller, R., Nyffenegger, T., and DeRobertis, E. M. (1983). *Cell*, **32**, 425.
24. Brimacombe R., Stiege, W., Kyriatsoulis, A., and Maly, P. (1988). In *Methods in enzymology*, (ed. H. F. Noller and K. Moldave), Vol. 164, pp. 287–309. Academic Press, San Diego.
25. Smith, K. C. (1976). *Photochem. Photobiol. Nucl. Acids*, **2**, 187.
26. Woppmann, A., Rinke, J., and Lührmann, R. (1988). *Nucleic Acids Res.*, **16**, 10 985.
27. Stiege, W., Glotz, C., and Brimacombe, R. (1983). *Nucleic Acids Res.*, **11**, 1687.
28. de Wachter, R., Maniloff, J. and Fiers, W. (1990). In *Gel electrophoresis of nucleic acids: a practical approach*. (ed. D. Rickwood, B. D. Hames), pp ● ●. IRL Press, Oxford.
29. Liautard, J. P., Sri-Widada, J., Brunel, C., and Jeanteur, P. (1982). *J. Mol. Biol.*, **162**, 623.
30. Lake, J. A. (1976). *J. Mol. Biol.*, **105**, 131.
31. Tischendorf, G. W., Zeichardt, H., and Stöffler, G. (1974). *Mol. Gen. Genet.*, **137**, 187.
32. Kastner, B and Lührmann, R. (1989). *EMBO J.*, **8**, 277.
33. Kastner, B., Bach, M., and Lührmann, R. (1990). *Proc. Natl Acad. Sci. USA*, **87**, 1710.
34. Kastner, B., Bach, M., and Lührmann, R. (1991). *J. Cell Biol.*, **112**, 1065.
35. Kastner, B., Kornstädt, U., Bach, M., and Lührmann, R. (1992). *J. Cell Biol.*, **116**, 839.
36. Lomant, A. J. and Fairbanks, G. (1976). *J. Mol. Biol.*, **104**, 243.
37. Stöffler-Meilicke, M. and Stöffler, G. (1988). In *Methods in enzymology*, (ed. H. F. Noller and K. Moldave), Vol. 164, pp. 503–520. Academic Press, San Diego.

6

Analysis of pre-mRNA splicing in yeast

ANDREW NEWMAN

1. Introduction

Many simple but powerful techniques have been developed to investigate gene expression in *Saccharomyces cerevisiae*, and a number of these have been employed in the analysis of RNA splicing. Test substrates can be transcribed from expression cassettes introduced into yeast by transformation or transplacement, and splicing can be monitored by RNA analysis or assay of a suitable reporter gene product. Gene disruption and inducible expression systems have been invaluable for investigating the roles of components of the splicing machinery in yeast. Many of the genes for yeast splicing factors have recently been isolated by molecular cloning, which is facilitated by the compact nature of the *Saccharomyces* genome and the fact that these genes are present as single copies. Splicing of yeast mRNA precursors can also be studied *in vitro*, since for *Saccharomyces* a simple method has been developed for making cell-free extracts capable of splicing synthetic pre-mRNA substrates.

Such studies have shown that the splicing pathway in yeast is very similar to that of higher eukaryotes and that there is considerable conservation of structure and function between splicing factors in yeast and mammalian cells. However, yeast is an extremely valuable system for the detailed analysis of splicing since it allows some approaches which are not possible with mammalian cells, particularly the genetic selection and isolation of splicing mutants.

2. Expression of exogenous genes

The approach described in this section is useful for the analysis of the effects of mutations in *cis*-acting intron sequences and for monitoring the effects of mutations in *trans*-acting factors on processing of specific pre-mRNAs.

2.1 Vectors

An expression cassette based on the yeast *ADH*1 gene transcription control regions is ideal for expression of test pre-mRNA substrates *in vivo* (1). The

6: Analysis of pre-mRNA splicing in yeast

ADH1 promoter achieves a high level of expression and has only two major start sites (10 bases apart) which facilitates RNA analysis. An intron-containing yeast gene fragment inserted downstream of this promoter is efficiently transcribed and is polyadenylated using the *ADH*1 gene 3' end formation signals. The entire expression cassette can be moved about using flanking *Bam*HI sites. For plasmid-based expression experiments, it is convenient if the cassette is carried on a centromere-based low copy number shuttle vector which also carries a single-strand DNA packaging origin to facilitate mutagenesis and sequencing of the substrate sequences (2).

2.2 Transformation of yeast

A procedure suitable for transforming yeast is given in *Protocol 1*.

Protocol 1. Yeast transformation

Reagents

- Appropriate strain of *S. cerevisiae*, e.g. trp1 URA3
- TE buffer (10 mM Tris–HCl, pH 7.4, 1 mM EDTA), sterilized by filtration through a 0.2 μm membrane filter
- DNA for transformation (1–5 μg in 1–10 μl of TE buffer)
- YEPD medium (1% yeast extract, 2% peptone, 2% dextrose), sterilized by autoclaving
- LiTE buffer (TE buffer containing 0.1 M lithium acetate), freshly made and sterilized by filtration through a 0.2 μm membrane filter
- PEG–TE buffer (TE buffer containing 40% PEG 4000)

Method

1. Grow the yeast cells in YEPD medium with shaking at 30 °C, to an $OD_{600nm} = 0.5–1.0$.
2. Centrifuge the culture (3 min at 2000 g) and discard the supernatant.
3. Resuspend the cell pellet in 1 vol. of sterile water and repeat step 2.
4. Wash the cells by resuspending them in 1 vol. of LiTE buffer and repeat step 2.
5. To render the cells competent for transformation, resuspend the cells in 0.25 vol. of LiTE buffer and incubate the suspension with gentle shaking at 30 °C for 60 min.
6. Harvest the cells by centrifugation (3 min at 2000 g) and resuspend them in 0.01 vol. of LiTE buffer. The competent cells are ready to use but can be stored at 4 °C for up to 24 h.
7. For transformation with DNA, mix 25 μl of the competent cell suspension with the transforming DNA (1–5 μg in 1–10 μl TE buffer) in a microcentrifuge tube and incubate this at 30 °C for 30 min.
8. Resuspend the cells by gentle agitation. Add 300 μl of PEG–TE buffer and mix by gentle agitation.

Continued

Protocol 1. *Continued*

9. Incubate the cells at 30 °C for 30 min and then heat shock them at 42 °C for 5 min.
10. Harvest the cells (10 sec in a microcentrifuge) and remove the supernatant by aspiration.
11. Add 100 µl of sterile TE buffer and resuspend the cells by gentle agitation.
12. Spread the cells on selective plates and incubate the plates at 30 °C for 48 h. Transformants usually appear after 18–24 h.

When the expression cassette is to be transplaced at a specific locus in the yeast genome, it is necessary to embed it in flanking DNA sequences from the locus of choice. The yeast *URA3* gene is particularly convenient for this purpose. Expression cassettes are inserted at the single *Stu*1 or *Eco*RV site in *URA3* to disrupt the *URA3* coding sequence. Restriction enzyme cleavage in the flanking *URA3* sequences yields a fragment which can be targeted to the *URA3* locus of a suitable recipient strain by homologous recombination (3). Clones carrying the transplacement are easily selected as ura3⁻ cells by virtue of their resistance to the 'suicide' substrate 5-fluoro-orotic acid (5FOA) (4). Transplacement of the expression cassette has the advantages that no selection is required thereafter and the copy number is absolutely fixed. This procedure is described in *Protocol 2*. The structure of the transplaced *ura3* region may be confirmed by PCR or Southern transfer analysis of individual 5FOA-resistant clones.

Protocol 2. Transplacement of mutant alleles at the *URA3* locus

Reagents

- 2 × 5FOA medium: add 0.7 g yeast nitrogen base (Difco), 100 mg 5-fluoro-orotic acid, 5 mg uracil, and 2 g glucose to 100 ml of distilled water; heat the mixture to 65 °C to dissolve the ingredients; sterilize the solution by filtration through a 0.2 µm membrane filter.

- 4% agar (Difco) sterilized by autoclaving

- −trp plates: dissolve 8 g Difco yeast nitrogen base (without amino acids), 55 mg adenine sulphate, 55 mg tyrosine, 55 mg uracil, 22 g agar and 11 g Difco casamino acids vitamin assay in 1 litre H₂O, autoclave and then add 100 ml of sterile 20% glucose and 20 ml of filter-sterilized 0.5% leucine)

- Competent cells of *trp*1 *URA3 S. cerevisiae* (see *Protocol 1*)

- *ura3* targeting DNA (1 µg/µl in TE buffer) linearized with a suitable restriction enzyme

- Plasmid DNA (1 µg/µl in TE buffer). Use a low copy number shuttle plasmid carrying the *TRP1* gene (*TRP1* acts as a co-transformation marker)

- 1 × 5FOA medium (1 vol. of 2 × 5FOA medium plus 1 vol. sterile distilled water)

- 5FOA plates: mix 100 ml of 2 × 5FOA medium with 100 ml of 4% agar at 65 °C. Pour the 5FOA agar into Petri dishes (8 ml per 50 mm plate) and allow it to set.

Continued

6: *Analysis of pre-mRNA splicing in yeast*

Protocol 2. *Continued*

Method

1. Mix 25 µl of competent cells with 2 µg of linearized targeting DNA and 1 µg of plasmid DNA.
2. Transform the cells and spread them on −trp plates as described in *Protocol 1*, steps 7–12. Incubate the plates at 30 °C for 48 h.
3. Pool the transformants (500–5000 colonies) by resuspending them in 2 ml of 1 × 5FOA medium and spread 20 µl of the cell suspension on a 50 mm 5FOA plate.
4. Incubate the 5FOA plates at 30 °C for 48 h to allow growth of cells carrying the transplaced disrupted *ura3* gene fragment.

3. Analysis of yeast RNA

3.1 Isolation of RNA from yeast

The best of various methods available for isolating RNA from yeast involves disruption of vegetative cells (haploid or diploid) with hot phenol in the presence of SDS, as described in *Protocol 3*. This releases total RNA efficiently, but the majority of the genomic DNA is retained within the cell wall. A 25 ml culture (see *Protocol 3*) will yield sufficient RNA for 10–20 primer extension reactions (see Section 3.2).

Protocol 3. Isolation of total RNA from yeast

Reagents

- Sodium acetate–EDTA buffer (50 mM sodium acetate, pH 5.3, 10 mM EDTA)
- Phenol equilibrated against sodium acetate–EDTA buffer
- Phenol : chloroform (1 : 1 v/v) containing 0.5% 8-hydroxyquinoline and equilibrated against 10 mM sodium acetate, pH 6.0, 0.1 M NaCl, 1 mM EDTA
- Appropriate strain of *S. cerevisiae* grown at 30 °C in 25 ml of YEPD medium (see *Protocol 1*) or suitable growth-selective medium to $OD_{600\,nm} = 1.0$–2.0.
- 10% SDS
- Chloroform : isoamyl alcohol (24 : 1 v/v)
- 3.0 M sodium acetate buffer, pH 5.3

Method

1. Harvest the cells by centrifugation (5 min at 2000 *g*).
2. Resuspend the cell pellet in 2 ml of sodium acetate–EDTA buffer. Add 0.25 ml 10% SDS and 2.5 ml of phenol.
3. Vortex (maximum speed) the mixture for 10 sec and then incubate it at 65 °C for 1 min. Repeat the vortexing and 65 °C treatment over a period of 5 min and then put the mixture on ice for 10 min.

Continued

Protocol 3. *Continued*

4. Separate the phases by centrifugation for 5 min at 2000 g and transfer the aqueous (upper) phase to a new tube.
5. Add 2.5 ml of phenol:chloroform and mix the phases vigorously (10 min on a gyrotory platform at maximum speed).
6. Separate the phases by centrifugation (step 4) and transfer the aqueous phase to a new tube. Add 2 ml of chloroform:isoamyl alcohol (24:1) and mix the phases vigorously as in step 5.
7. Separate the phases by centrifugation (step 4) and transfer the aqueous phase to a new tube. Add 0.1 vol. of 0.3 M sodium acetate, pH 5.3, and 2.5 vol. of absolute ethanol.
8. Mix the contents of the tube by inversion and allow the RNA to precipitate at $-20\ °C$ for 1 h.
9. Recover the total RNA by centrifugation (10 min at 2500 g, 4 °C) and redissolve it in 1.0 ml of sterile distilled water. Store the RNA in aliquots at $-70\ °C$.

Total RNA prepared as described in *Protocol 3* can be analysed by primer extension (see Section 3.2). Alternatively, nuclease S1 analysis or Northern transfer (Chapter 2, Section 3) can be used. For low abundance species and for the characterization of lariat splicing intermediates it is advantageous first to purify poly(A)$^+$ RNA from the total RNA by selection with Hybond™-mAP (see Chapter 1, Section 3.3) or by chromatography on oligo(dT)-cellulose as described in *Protocol 4*.

Protocol 4. Isolation of poly(A)$^+$ RNA from yeast total RNA

Reagents

- Loading buffer (1.0 M LiCl, 20 mM Tris-HCl, pH 7.4, 2 mM EDTA, 0.2% SDS)
- Washing buffer (0.5 M LiCl, 10 mM Tris-HCl, pH 7.4, 1 mM EDTA, 0.1% SDS)
- Elution buffer (10 mM Tris-HCl, pH 7.4, 1 mM EDTA, 0.1% SDS)
- 3.0 M sodium acetate buffer, pH 5.3
- Glycogen, nuclease-free (Sigma, 10 mg/ml)
- Oligo(dT)-cellulose (Collaborative Research); 0.1 ml bed vol. in a 1 ml disposable plastic syringe
- Total yeast RNA (1-2 mg/ml in water) prepared as described in *Protocol 3*

Method

1. Pre-equilibrate the oligo(dT)-cellulose column in washing buffer.
2. Denature the RNA at 65 °C for 5 min and then cool it quickly to 20 °C.
3. Add 1.0 ml of loading buffer to the RNA and load the solution on to the oligo(dT)-cellulose column.
4. Wash the column with 5 ml of washing buffer.

Continued

6: Analysis of pre-mRNA splicing in yeast

Protocol 4. *Continued*

5. Elute the poly(A)⁺ RNA from the column with three 0.4 ml aliquots of elution buffer.
6. Pool the eluted fractions and precipitate the RNA by addition of 0.1 vol. of 3 M sodium acetate buffer, pH 5.3, 10 µg of nuclease-free glycogen (to act as carrier), and 2.5 vol. of absolute ethanol. Leave the mixture for at least 1 h at −20 °C.
7. Recover the precipitated poly(A)⁺ RNA by centrifugation (10 min at 2500 g, 4 °C) and decant the supernatant.
8. Remove any remaining drops of supernatant by aspiration and dissolve the RNA pellet in 25 µl of sterile distilled water. Store the RNA at −20 °C.

3.2 RNA analysis by primer extension

This is, in many ways, the method of choice for determining the steady-state levels of substrates, processing intermediates, and mRNA products in yeast poly(A)⁺ RNA. It is particularly useful for analysing lariat intermediates, which generate characteristic primer extension products terminating at the site of the branch (see *Figure 1*). A 5′ end-labelled oligodeoxyribonucleotide primer is used (typically 20–25 nt long). The primer can be designed so that it is specific for RNA derived from the expression cassette, or so that endogenous RNA acts as an internal standard. Primer extension analysis using reverse transcriptase is described in *Protocol 5*.

Protocol 5. Primer extension analysis using reverse transcriptase

Equipment and reagents

- Poly(A)⁺ RNA prepared as described in Protocol 4
- RNasin (Promega, 40 units/µl)
- AMV reverse transcriptase (Anglian Biotec, 20 units/µl)
- 0.5 M DTT stock solution
- dNTP stock solution (1 mM each of dATP, dCTP, dGTP, and dTTP)
- 10 mg/ml tRNA stock solution (Sigma)
- Appropriate oligodeoxyribonucleotide primer (1 ng/µl in water), 5′ end-labelled with [γ-³²P]ATP and T4 polynucleotide kinase as described in Chapter 2, Protocol 2
- Hybridization buffer (50 mM KCl, 20 mM Tris–HCl, pH 8.5, 0.5 mM EDTA, 8 mM MgCl$_2$)
- Extension mixture: (Using stock solutions make up the following mixture in sterile water) 0.5 unit/µl RNasin, 0.25 unit/µl AMV reverse transcriptase, 5 mM DTT, 1 mM each of dATP, dGTP, dCTP, and dTTP
- Stop cocktail (0.3 M sodium acetate, pH 5.3, 1 mM EDTA, 0.1% SDS, 25 µg/ml tRNA)
- Phenol solution equilibrated against stop cocktail
- Materials for denaturing polyacrylamide/urea gel electrophoresis (Chapter 1, Protocol 8)

Continued

Protocol 5. *Continued*

Method

1. Hybridize 0.5 µg of poly(A)+ RNA with 1 ng of 5' end-labelled primer at 37 °C for 30 min in 10 µl of hybridization buffer.
2. Add 10 µl of extension mixture and incubate the reaction at 41 °C for 30 min.
3. Stop the reaction by the addition of 100 µl of stop cocktail. Add 1 vol. of phenol solution and vortex the reaction (maximum speed) for 20 sec.
4. Separate the phases by centrifugation for 1 min in a microcentrifuge and remove the aqueous phase to a fresh tube.
5. Precipitate the nucleic acids from the aqueous phase by adding 2.5 vol. of ethanol. Leave the mixture for 1 h at −20 °C.
6. Recover the nucleic acids by centrifugation (3 min in a microcentrifuge) and remove the supernatant by aspiration.
7. Redissolve the nucleic acid pellet in 3 µl of formamide–dyes loading buffer and fractionate the labelled cDNA products by denaturing polyacrylamide gel electrophoresis as described in Chapter 1, *Protocol 8*.

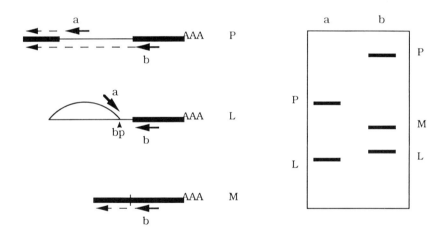

Figure 1. Analysis of pre-mRNA splicing by primer extension. Left panel, structures of the pre-mRNA, lariat intermediate and spliced mRNA product. Intron sequences are shown as light lines, and exon sequences as heavy lines. The oligonucleotide primers *a* and *b* are specific for intron and exon sequences, respectively. Dashed lines indicate the extension of the primer by reverse transcriptase, which is arrested when either the 5'-end of the RNA or a branchpoint is reached. AAA represents the poly(A) tail, and bp marks the branchpoint in the lariat intermediate. Right panel, Cartoon of a gel analysis of the cDNA products from reverse transcription of poly(A)+ RNA using primers *a* and *b*. P, pre-mRNA; L, lariat intermediate; M, mRNA.

An important consideration relevant to the interpretation of results from the analysis of splicing *in vivo* is that generally splicing is not rate-limiting for mRNA production in the living cell. The splicing efficiency of wild-type genes can be many times greater than that required to maintain the normal level of mRNA. Consequently, mutations (for example, in splice-site sequences) often have no effect on mRNA levels even though they have drastic effects on splicing efficiency. It is also important to consider the level of pre-mRNA; this can rise dramatically as a result of intron mutations which nevertheless fail to affect mRNA levels. The effects of intron mutations on splicing efficiency are, therefore, best described by the ratio of pre-mRNA to mRNA (5).

Fractionation of the cDNA products of primer extension reactions by denaturing polyacrylamide gel electrophoresis often allows the products and intermediates of splicing to be identified unambiguously. The assignment of these products can be facilitated by co-electrophoresis of dideoxynucleotide sequencing size markers from reactions set up using the same oligonucleotide primer and a single-stranded template from the gene of interest.

Primer extension analysis of splicing sometimes produces cDNA products which cannot be identified by electrophoretic mobility alone. If this happens it may be necessary to sequence the cDNA products directly to determine their identity. In this case the cDNA must first be amplified by PCR using appropriate forward and reverse primers (see Chapter 2, Section 3.4 and Chapter 3, Section 2.2).

4. Genetic strategies for investigating pre-mRNA splicing

One of the advantages of using yeast as an experimental organism is the availability of a number of sophisticated genetic approaches which have been developed in the course of studies of gene expression and other fundamental processes. These are particularly useful for the dissection of complex systems such as the pre-mRNA splicing machinery, where biochemical approaches to the identities and functions of the components are technically difficult.

4.1 Isolation and characterization of *PRP* genes and their products

The genes of the *PRP* (Precursor RNA Processing) family, which currently number more than thirty, encode protein components of the splicing machinery. The great majority of *prp* gene alleles have been isolated using temperature-sensitive lethal mutations which cause the accumulation of pre-mRNAs or lariat splicing intermediates. In wild-type strains introns are efficiently removed by splicing and rapidly degraded so that the steady-state levels of intron sequences are very low. The isolation and analysis of *prp* gene mutants involves screening libraries of temperature-sensitive (*ts*) lethal strains for accumulation of intron sequences at the restrictive temperature. In one such study about 1000 *ts* strains were isolated after ethylmethanesulphonate mutagenesis, followed by replica plating on rich medium and screening individual colonies able to grow at 23 °C but not at 37 °C (6). The *ts* strains were then screened for accumulation of intron-containing RNA species after growth for 2 h at 37 °C. Total RNA was fractionated by electrophoresis on formaldehyde–agarose gels and transferred by electroblotting to Genescreen membranes. Filters were probed with a labelled intron fragment of the actin gene to detect pre-mRNA, the lariat intermediate, and intron species, then stripped and reprobed with an actin exon fragment to detect mRNA and pre-mRNA. RNA from candidate *prp* mutant isolates was then examined further by primer extension analysis, and selected clones were subjected to complementation and segregation analysis (6).

New *prp* mutations are extremely useful since they allow the cognate wild-type *PRP* gene to be isolated from a library of genomic DNA fragments from a wild-type strain by complementation of the *ts* lethality phenotype. Many of these genes have now been cloned (reviewed in ref. 7), and several have been assigned to specific spliceosomal snRNPs using antisera raised against the protein made in bacteria. Antisera raised against *prp* proteins are invaluable reagents for the analysis of splicing *in vitro* (8). Sequence analysis of *PRP* genes has revealed some features of interest (for example, homologies with RNA helicases and zinc finger motifs) which may provide clues about the functions of these proteins in splicing.

4.2 Genetic approaches to splicing based on interactive suppression

The rationale behind using interactive suppression is that mutations (in the pre-mRNA or in a splicing factor) which modify or disrupt important interactions between spliceosome components can sometimes be suppressed by a compensating mutation in another factor which interacts with the molecule carrying the first mutation. This method can be a powerful way of identifying functionally-important interactions within the spliceosome. One of its strengths is that it can be focused on particular aspects of recognition or catalysis and, therefore, yield information about the precise functions of the splicing factors involved in the suppression events.

Interactive suppression has been used to identify genes whose products interact with the intron branchpoint sequence and with the 5′ splice site. The method involves setting up a transcription unit in which expression of a reporter gene demands the prior removal of an upstream intron. The reporter gene product can be an enzyme such as β-galactosidase, which can be assayed using a chromogenic substrate. Alternatively, one can employ a gene such as *HIS*4, whose expression allows the growth of *his*4 strains on histidinol, or the bacterial neomycin-resistance gene (*NEO*), which confers resistance to the aminoglycoside antibiotic, G418.

Introduction of a mutation into, for example, one of the splice sites in such a reporter gene fusion, blocks splicing and so prevents expression of reporter gene activity. It is then possible to search for extragenic suppressors after mutagenesis of a strain carrying the gene fusion. It is advantageous to carry out these suppressor screens in a diploid background since suppressors which arise are then very likely to be dominant, and can include recessive lethal alleles which could not appear in haploid screens. Candidate suppressors are challenged with a panel of substrates carrying various intron mutations both to establish allele-specificity and to confirm that the suppressor acts in *trans* at the level of splicing. Dominant allele-specific interactive suppressors identified in this way can be cloned by virtue of their suppressor phenotype. A library of genomic DNA fragments is made from the strain carrying the suppressor using partial digestion with *Sau*3AI to 10–20 kbp fragments. These are then cloned into the *Bam*HI site of a low copy number centromere-based shuttle vector (9). Plasmids which carry the suppressor can then be isolated after transformation of a strain harbouring the reporter gene fusion. *Table 1* summarizes the main steps in a research project designed to isolate the gene for a component of the splicing machinery and to investigate its role in splicing.

This sort of approach has been used to isolate the *prp*16-1 suppressor gene (involved in intron branchpoint recognition) (10, 11) and an allele of the *snr*7 gene (encoding U5 snRNA) which suppressed a mutation in the G1 nucleotide at the 5' splice site (12). Interactive suppression has also been used successfully to isolate suppressors of mutations in snRNA and *prp* genes. For example, a mutation in U4 snRNA, which destabilizes the U4/U6 base-pairing interaction, results in a cold-sensitive phenotype. This was used to isolate an interactive suppressor which restored cold-insensitive growth. The suppressor proved to be an allele of the *PRP*24 gene, previously shown to encode an essential splicing factor (13).

4.3 Functional analysis of splicing factors *in vivo*

Once isolated by molecular cloning, genes encoding components of the splicing machinery can be manipulated *in vitro*. Simple techniques are available to assess the effects of manipulation on gene function *in vivo*. First a diploid strain carrying only one functional copy of the gene is made by gene disruption. This involves replacing most of the gene sequences with a marker gene (such as *LEU*2) for which a selection is available, and targeting the disrupted gene to one of the chromosomal loci by homologous recombination (3). If the gene is essential, sporulation of such a strain will produce only two viable spores per tetrad, and the survivors will not carry the disrupting marker.

Viable haploids carrying a gene disruption can be isolated if they are provided with a functional copy of the gene of interest on a single copy plasmid. New alleles of the gene, created *in vitro*, can then be introduced on a second

Table 1. Stages in an interactive suppression study of pre-mRNA splicing

1. Construct a diploid strain carrying *lacZ* and *NEO* fusion genes transplaced at the *ura*3 loci.
2. The inclusion of an intron mutation (e.g. a 5' splice-site mutation) in each fusion blocks splicing, preventing expression of reporter gene products.
3. Mutagenize the cells with UV irradiation.
4. Screen for β-galactosidase expression using X-Gal, and select blue colonies.
5. Rescreen the colonies for G418-resistance.
6. Isolate poly(A)$^+$ RNA (see *Protocol 4*).
7. Analyse the poly(A)$^+$ RNA by primer extension (see *Protocol 5*) using *lacZ* and *NEO* primers.
8. Introduce an *ADH1* cassette expressing pre-mRNA with the intron mutation and other mutations to examine allele-specificity.
9. Isolate DNA from the mutant strain and subject it to partial restriction enzyme digestion with *Sau*3AI.
10. Construct a genomic library from the *Sau*3AI digest in a centrometric (low copy number) shuttle vector.
11. Transform a reporter strain carrying the original *lacZ* reporter fusion with the genomic library and screen the transformants for β-galactosidase expression using X-Gal.
12. Grow up blue colonies and isolate the plasmid DNA.
13. Prepare a restriction map of the complementing sequence by sub-cloning.
14. Sequence the smallest complementing fragment.
15. Compare the sequence obtained with a sequence database to identify the complementing gene or to reveal homologies with other genes.
16. Construct a strain carrying a disruption in the complementing gene to see whether it is an essential gene.
17. Transform a suitable bacterial host with the yeast gene in an appropriate expression vector.
18. Overproduce the recombinant protein in bacteria and purify the protein.
19. Raise antisera against the recombinant protein.
20. Use the antisera to determine whether the yeast factor is a snRNP protein or spliceosome component.

plasmid using a different selective marker. If the first plasmid carries a *URA3* or *LYS2* marker, the wild-type gene can then be evicted by plating the cells on medium containing a 'suicide' substrate (5-fluoro-orotic acid or α-aminoadipate, respectively) (4, 14). This allows the mutant alleles to be studied as the sole source of the gene product of interest. Experiments of this type have been very informative in defining functionally important regions of snRNA and protein splicing factors (15–18).

5. Analysis of RNA splicing in yeast extracts

Analysis of pre-mRNA splicing reactions *in vitro* has shown that splicing involves two successive *trans*-esterification reactions and generates branched 'lariat' molecules as intermediates and products. These reactions occur in a large particle, the spliceosome, into which the pre-mRNA is assembled by the ordered interaction of multiple RNA and protein splicing factors. Some of the methods of particular relevance to yeast are discussed in this section.

5.1 Preparation of active extracts from yeast

An extract of yeast active in splicing was originally developed (19) using a protease-deficient strain (EJ101:*MAT*a *trp*1 *pro*1-126 *prb*1-112 *pep*4-3 *prc*1-126) but it has since proved possible to make active extracts from a variety of other strains, including those with no protease deficiencies. *Protocol 6* describes the procedure.

Protocol 6. Preparation of a splicing extract from yeast

Equipment and reagents

- YEPD medium (see *Protocol 1*)
- SB30 buffer (50 mM Tris–HCl, pH 8.0, 10 mM $MgCl_2$, 1 M sorbitol, 30 mM DTT)
- SB3 buffer (50 mM Tris–HCl, pH 8.0, 10 mM $MgCl_2$, 1 M sorbitol, 3 mM DTT)
- Zymolyase solution. Dissolve zymolyase-100T (Seikagoku Kogyo) at 20 mg/ml in 50 mM potassium phosphate buffer, pH 7.5
- Buffer A (10 mM Hepes–KOH, pH 7.9, 10 mM KCl, 0.5 mM DTT, 1.5 mM $MgCl_2$)
- 2.0 M KCl
- Buffer D (20 mM Hepes–KOH, pH 7.9, 0.2 mM EDTA, 0.5 mM DTT, 50 mM KCl, 20% glycerol)
- Dounce homogenizer (15 ml, all-glass) with a tight-fitting pestle
- Sorvall centrifuge with GS3 and SS34 rotors and tubes; Beckman ultracentrifuge with Ti60 rotor and tubes (or equivalents)

Method

1. Grow the yeast with shaking at 30 °C in YEPD medium (1 litre in a 5 litre flask) to an $OD_{600} = 2.0–4.0$.
2. Harvest the cells by centrifugation at 1200 g (Sorvall GS3 or equivalent) for 5 min and resuspend them in 15 ml of SB30 buffer. Incubate the cells at 25 °C without shaking for 15 min.
3. Collect the cells by centrifugation (as in step 2) and resuspend them in 15 ml of SB3 buffer.
4. Add 100 µl of zymolyase solution and shake the cells gently at 30 °C for 40 min to form spheroplasts. *Carry out subsequent steps at 0–4 °C.*
5. Collect the spheroplasts by centrifugation at 1200 g for 5 min and wash them once by gentle resuspension in 15 ml of SB3 buffer. Collect them by centrifugation as before.

Continued

Protocol 6. *Continued*

6. Gently resuspend the spheroplasts in 8 ml of buffer A and lyse them by homogenization in a Dounce homogenizer using 5 strokes with a tight-fitting pestle.
7. Add 2.0 M KCl to the lysate to a final concentration of 0.2 M and stir the mixture gently on ice for 30 min.
8. Remove cell debris by centrifugation at 26 000 g for 30 min (Sorvall SS34 rotor or equivalent).
9. Centrifuge the supernatant at 100 000 g for 60 min (Beckman Ti60 rotor or equivalent) and carefully collect the upper clear part of the supernatant.
10. Dialyse the clear supernatant for 3 h against 1 litre of buffer D and spin the dialysed supernatant at 26 000 g for 20 min to remove insoluble material.
11. Flash-freeze the supernatant in small aliquots (50 µl) in liquid nitrogen. Store them at −70 °C. Extracts are stable for at least 2 years at −70 °C. A typical protein concentration is 20–30 mg/ml.

5.2 Splicing reactions *in vitro*

Synthesize labelled pre-mRNA substrates for splicing reactions as run-off transcripts using T7 RNA polymerase and [α-^{32}P]UTP (specific activity, 20 Ci/mmol) (Chapter 1, *Protocol 4*). It is important to purify the substrate by electrophoresis on a 6% polyacrylamide/8 M urea gel followed by elution (Chapter 1, *Protocol 8*) in order to separate the full-length pre-mRNA from degradation products and prematurely-terminated transcripts.

The procedure used for studying splicing reactions in extracts is described in *Protocol 7*. The pre-mRNA substrate, splicing intermediates and mRNA and intron products are separated by denaturing gel electrophoresis. The polyacrylamide concentration must be optimized in pilot experiments for optimal separation of lariat products, which migrate anomalously slowly, from pre-mRNA and degradation products. The optimum concentration varies according to the relative sizes of the intron and exon components of the pre-mRNA.

Protocol 7. Splicing of pre-mRNA *in vitro* using yeast extracts

Equipment and reagents

- 5 × splice cocktail (10 mM ATP, 12.5 mM MgCl$_2$, 0.3 M potassium phosphate, pH 7.0)
- Splicing extract (see *Protocol 6*)
- 30% PEG (PEG 8000)
- ^{32}P-labelled pre-mRNA substrate (10 nM) labelled with [α-^{32}P]UTP at 20 Ci/mmol as described in Chapter 1, *Protocol 4*
- Stop solution (1 mg/ml proteinase K, 50 mM EDTA, 1% SDS)
- Diluent (50 mM sodium acetate, pH 5.3, 1 mM EDTA, 0.1% SDS, 25 µg/ml *E. coli* tRNA)

Continued

6: Analysis of pre-mRNA splicing in yeast

Protocol 7. *Continued*

- Phenol:chloroform (1:1 v/v) equilibrated against diluent
- 3.0 M sodium acetate buffer, pH 5.3
- Materials for denaturing polyacrylamide/urea gel electrophoresis (Chapter 1, *Protocol 8*)

Method

1. Thaw a tube of splicing extract on ice.
2. Assemble a splicing reaction in a microcentrifuge tube on ice from the following components[a]:
   - splicing extract   8 µl
   - 5 × splice cocktail   4 µl
   - 30% PEG   2 µl
   - pre-mRNA substrate   vol. for 20 fmol
   - water   to 20 µl
3. Start the reaction by placing the tube at 23 °C.
4. Take samples (5 µl) at intervals (the intermediates and products of splicing are detectable after 5–10 min at 23 °C) into microcentrifuge tubes containing 1 µl of stop solution.
5. Incubate the samples at 37 °C for 20 min.
6. Add 200 µl of diluent to each sample and remove the proteins by vortexing with an equal vol. of phenol:chloroform. Separate the phases by centrifugation in a microcentrifuge for 1 min.
7. Remove the upper (aqueous) phase into a new tube.
8. Add 0.1 vol. of 3 M sodium acetate (pH 5.3) and 2.5 vol. of ethanol. Leave the mixture for 30 min at −20 °C.
9. Recover the precipitated RNA by centrifugation for 3 min in a microcentrifuge.
10. Rinse the RNA pellet with 70% ethanol, aspirate the ethanol to allow the RNA to air dry.
11. Redissolve the RNA in 3 µl of formamide–dyes loading buffer and fractionate the RNA on a polyacrylamide/8 M urea sequencing gel (Chapter 1, *Protocol 8*).

[a] Final concentrations in the reaction are: 2 mM ATP, 2.5 mM $MgCl_2$, 3% PEG 8000, 60 mM potassium phosphate, 20 mM KCl, 8 mM Hepes, 8% glycerol, 80 µM EDTA, 0.2 mM DTT, 1 nM pre-mRNA.

Splicing reactions of the sort described in *Protocol 7* have been invaluable in the analysis of the RNA intermediates and products of splicing. It is also possible to monitor the formation of splicing complexes using non-denaturing gel electrophoresis, for which several methods have been published (20–22). Extracts from strains carrying *ts prp* mutations are specifically temperature-sensitive *in vitro*. Such inactivated extracts can be used in complementation experiments with extracts from other *prp* mutants or with purified and partially purified splicing factors (see Chapter 3, Section 4). It has also been possible to complement a heat-inactivated extract (from a *prp*11 strain) with *PRP*11 protein made in a reticulocyte lysate translation system programmed with *PRP*11 mRNA synthesized by T7 polymerase transcription (23). This approach allows ^{35}S-labelled or epitope-tagged protein to be followed into the spliceosome.

5.3 Depletion of components from splicing extracts

Splicing extracts may be depleted of specific components essential for the splicing reactions using biochemical or genetic approaches. Complementation of these depleted extracts allows the activity of specific factors to be analysed.

5.3.1 Targeted destruction of snRNAs in yeast splicing extracts

This approach has been particularly informative in investigations of U2 and U6 snRNP function. The technique is based on the use of specific synthetic oligonucleotides to target endogenous U2 or U6 snRNA. The oligonucleotide–RNA hybrid is attacked by RNase H activity which is present in extracts, thereby abolishing splicing activity (24, 25). The oligonucleotide is itself rapidly degraded. The depleted extract can only be restored to activity by the addition of synthetic U2 or U6 RNA made by T7 RNA polymerase. Restoration of splicing activity via assembly of the synthetic snRNA into snRNPs is monitored by the subsequent addition of suitable synthetic pre-mRNA substrates. Such depleted extracts allow the effects of mutations in the complementing snRNA to be assayed biochemically. The technique for depleting extracts using oligonucleotide-mediated RNase H action is described in *Protocol 8*.

Protocol 8. Oligonucleotide-mediated RNase H cleavage of snRNA and reconstitution of splicing activity

Equipment and reagents

- 5 × RNase H/splice cocktail (10 mM ATP, 12.5 mM MgCl$_2$, 0.3 M potassium phosphate, pH 7.0, 10 mM spermidine)
- Splicing extract (see *Protocol 6*)
- 30% PEG, stop solution, diluent, phenol: chloroform, 3 M sodium acetate buffer, pH 5.3, and ^{32}P-labelled pre-mRNA substrate (see *Protocol 7*)
- Synthetic oligonucleotide complementary to an exposed portion of the target snRNA[a]
- U2 or U6 snRNA (50 nM) synthesized by *in vitro* transcription as described in Chapter 1, *Protocol 3*

Method

1. Thaw a tube of splicing extract on ice.
2. Assemble a splicing reaction in a microcentrifuge tube on ice from the following components:
   - splicing extract   4 µl
   - 5 × RNase H/splice cocktail   2 µl

Continued

Protocol 8. Continued

- 30% PEG 1 μl
- oligonucleotide vol. for 3 nmol
- water to 10 μl

3. Incubate the mixture at 30 °C for 30 min to allow RNase H action to occur.
4. Add 1 μl of synthetic U2 or U6 snRNA to 4 μl of the depleted extract and incubate at 23 °C for 10 min to allow snRNP assembly.
5. Add synthetic ^{32}P-labelled pre-mRNA (1 fmol) and incubate the extract at 23 °C for 30 min.
6. Add 1 μl of stop solution and analyse the RNA products as described in *Protocol 7*, steps 5–11.

^{a} See refs 24 and 25.

5.3.2 Genetic depletion of specific snRNAs

Oligonucleotide/RNase H-mediated inactivation is not always successful. U5 snRNP in particular has been intractable. 'Genetic depletion' has been used successfully to produce extracts depleted specifically for U1, U2, or U5 snRNPs. The approach involves construction of yeast strains which contain a disrupted chromosomal copy of the snRNA gene of interest. The disruption is complemented by a plasmid containing a copy of the snRNA gene under the control of the *GAL*10 upstream activation region. The snRNA is expressed efficiently in the presence of galactose, but growth in glucose-containing media represses *gal*-mediated transcription. Extracts specifically lacking the snRNA can be prepared (as in *Protocol 6*) after growth for 16 h at 30 °C in YEPD medium containing 4% glucose. Such extracts are inactive in splicing assays and exhibit defects at different stages in the assembly of splicing complexes, according to the identity of the depleted snRNP (22, 26).

References

1. Ammerer, G. (1983). In *Methods in enzymology*, Vol. 101 (ed. R. Wu, L. Grossman, and K. Moldave), p. 192.
2. Sikorski, R. and Hieter, P. (1989). *Genetics*, **122**, 19.
3. Rothstein, R. (1983). In *Methods in enzymology*, Vol. 101 (ed. R. Wu, L. Grossman, and K. Moldave), p. 202.
4. Boeke, J. D., Trueheart J., Natsoulis, G., and Kink, G. R. (1987). *Methods in enzymology*, Vol. 154 (ed. R. Wu and L. Grossman), p. 164.
5. Pikielny, C. W. and Rosbash, M. (1985). *Cell*, **41**, 119.
6. Vijayraghavan, U., Company, M., and Abelson, J. (1989). *Genes Dev.*, **3**, 1206.
7. Guthrie, C. (1991). *Science*, **253**, 157.
8. Lossky, M., Anderson, G. J., Jackson, S. P., and Beggs, J. (1987). *Cell*, **51**, 1019.
9. Johnston, J. R. (1987). *Yeast: a practical approach*, (ed. I. Campbell and J. H. Duffus) p. 107–123. IRL Press, Oxford.

10. Couto, J. R., Tamm, J., Parker, R., and Guthrie, C. (1987). *Genes Dev.*, **1**, 445.
11. Burgess, S., Couto, J. R., and Guthrie, C. (1990). *Cell*, **60**, 705.
12. Newman, A. J. and Norman, C. (1991). *Cell*, **65**, 115.
13. Shannon, K. W. and Guthrie, C. (1991). *Genes Dev.*, **5**, 773.
14. Chattoo, B., Sherman, F., Fjellstedt, T., Menhnert, D., and Ogur, M. (1979). *Genetics*, **93**, 51.
15. Igel, A. H. and Ares, M. (1988). *Nature*, **334**, 450.
16. Shuster, E. O. and Guthrie, C. (1988). *Cell*, **55**, 41.
17. Seraphin, B., Kretzner, L., and Rosbash, M. (1988). *EMBO J.*, **7**, 2533.
18. Smith, V. and Barrell, B. G. (1991). *EMBO J.*, **10**, 2627.
19. Newman, A. J., Lin, R. J., Cheng, S., and Abelson, J. (1985). *Cell*, **42**, 335.
20. Cheng, S. and Abelson, J. (1987). *Genes Dev.*, **1**, 1014.
21. Legrain, P., Seraphin, B., and Rosbash, M. (1988). *Mol. Cell. Biol.*, **8**, 3755.
22. Seraphin, B. and Rosbash, M. (1989). *Cell*, **59**, 349.
23. Chang, T., Clark, M. W., Lustig, A. J., Cusick, M. E., and Abelson, J. (1988). *Mol. Cell. Biol.*, **8**, 2379.
24. Fabrizio, P. and Abelson, J. (1990). *Science*, **250**, 404.
25. McPheeters, D. S., Fabrizio, P., and Abelson, J. (1989). *Genes Dev.*, **3**, 2124.
26. Seraphin, B., Abovich, N., and Rosbash, M. (1991). *Nucleic Acids Res.*, **19**, 3857.

A1

Suppliers of specialist items

Aldrich Chemical Company Ltd, The Old Brickyard, New Road, Gillingham, Dorset SP5 4BR, UK.
Aldrich Chemical Company Inc., 940 West St. Pool Avenue, Millwaukee, WI 53233, USA.
American Type Culture Collection (ATCC), 12301 Park Lawn Drive, Rockville, MD 20852, USA.
Amersham International PLC, Lincoln Place, Green End, Aylesbury, Bucks. HP20 2TP, UK.
Amersham Corporation, 2636 South Clearbrook Drive, Arlington Heights, IL 60005, USA.
Amicon Ltd, Upper Mill, Stonehouse, Gloucester GL10 2BJ, UK.
Amicon Division, WR Grace & Co., 72 Cherryhill Drive, Beverley, MA 01915-1065, USA.
Anglian Biotec Ltd., Whitehall House, Whitehall Road, Colchester, Essex CO2 8HA, UK.
Applied Biosystems Inc., 850 Lincoln Center Drive, Foster City, CA 94404, USA.
Applied Biosystems Ltd., Kelvin Close, Birchwood Science Park North, Warrington WA3 7PB, UK.
Beckman Instruments UK Ltd., Progress Road, Sands Industrial Estate, High Wycombe, Bucks. HP12 4JL, UK.
Beckman Instruments Inc., PO Box 3100, 2500 Harbor Boulevard, Fullerton, CA 92634, USA.
Becton Dickinson Labware, 2 Bridgewater Lane, Lincoln Park, NJ 07035, USA.
Becton Dickinson UK Ltd., Between Towns Road, Cowley, Oxford OX4 3LY, UK.
Bethesda Research Laboratories (BRL); see Gibco/BRL.
Bio 101 Inc., PO Box 2284, La Jolla, CA 92038-2284, USA.
   c/o Stratech Scientific Ltd, 61–63 Dudley Street, Luton, Beds. LU2 0HP, UK.

Appendix 1

Bio-Rad Laboratories Ltd, Maylands Avenue, Hemel Hempstead, Herts HP2 7TD, UK.

Bio-Rad Laboratories, Division Headquarters, 3300 Regatta Boulevard, Richmond, CA 94804, USA.

Boehringer Mannheim GmbH Biochemica, PO Box 31 01 20, D-6800 Mannheim, Germany.

Boehringer Mannheim UK (Diagnostics/Biochemicals) Ltd, Bell Lane, Lewes, East Sussex BN7 1LG, UK.

Boehringer Mannheim Corporation, Biochemical Products, PO Box 50414, Indianapolis, IN 46250-0414, USA.

Branson Ultrasonic Corporation, 41 Eagle Road, Danbury, CT 06813, USA.
  c/o Lucas Dawes Ultrasonics Ltd, Concord Road, Western Avenue, London W3 0SD, UK.

BRL; see Gibco/BRL.

Buchler (Haake Buchler Instruments, Inc), 244 Saddle River Road, Saddle Brook, NJ 07662, USA.

Calbiochem, PO Box 12087, San Diego, CA 92112-4180, USA.

Calbiochem-Novabiochem (UK) Ltd, 3 Heathcoat Building, Highfields Science Park, University Boulevard, Nottingham NG7 2QJ, UK.

Cambridge Bioscience Ltd, 25 Signet Court, Stourbridge Common Business Centre, Swans Road, Cambridge CB5 8LA, UK.

Cambridge Research Biochemicals, Gadbrook Park, Northwich, Cheshire, UK.

Cambridge Research Biochemicals Inc., Wilmington, DE 19897, USA.

Cetus; see Perkin Elmer-Cetus.

Clontech Laboratories Inc., 4055 Fabian Way, Palo Alto, CA 94303, USA.
  c/o Cambridge Bioscience Ltd, 25 Signet Court, Stourbridge Common Business Centre, Swans Road, Cambridge CB5 8LA, UK.

Collaborative Research Inc., 128 Spring Street, Lexington, MA 02173, USA.
  c/o Universal Biologicals Ltd, 12-14 St Ann's Crescent, London SW18 2LS, UK.

Costar, One Alewife Center, Cambridge, MA02140, USA.

Costar (UK) Ltd, Victoria House, 28-38 Desborough Street, High Wycombe, Bucks. HP11 2NF, UK.

Cruachem Ltd, Todd Campus, West of Scotland Science Park, Acre Road, Glasgow G20 0UA, UK.

Difco Laboratories Ltd, Central Avenue, East Molesley, Surrey KT8 0SE, UK.

Difco Laboratories, PO Box 331058, Detroit, Michigan 48232-7058, USA.

Du Pont Co. (Biotechnology Systems Division), PO Box 80024, Wilmington, DE 19880-0024, USA.

Du Pont UK Ltd, Wedgwood Way, Stevenage, Herts SG1 4QN, UK.

Falcon; contact Becton Dickinson.

5 Prime-3 Prime, Inc., 5603 Arapahoe, Boulder, CO 80303, USA.
  c/o CP Laboratories, PO Box 22, Bishop's Stortford, Herts CM23 3DX, UK.

Appendix 1

Flow (ICN Flow), Eagle House, Peregrine Business Park, Gomm Road, High Wycombe HP13 7DL, UK.
Flow (ICN Biomedical Inc.), 3300 Highland Avenue, Costa Mesa, CA 92626, USA.
Fluka Chemie AG, Industriestrasse 25, CH-9470 Buchs, Switzerland.
Fluka Chemicals Ltd, The Old Brickyard, New Road, Gillingham, Dorset SP5 4BR, UK.
Gibco BRL (Life Technologies Ltd), Trident House, Renfrew Road, Paisley PA3 4EF, UK.
Gibco BRL (Life Technologies Inc.), 3175 Staler Road, Grand Island, NY 14072-0068, USA.
Glen Research, 44901 Falcon Place, Sterling, VA 22170, USA.
  c/o Spectra Ltd., 2–4 Wigton Gardens, Stanmore, Middx. HA7 1BG, UK.
Greiner GmbH, Maybachstrasse 2, D-7443 Frickenhausen, Germany.
  c/o Philip Harris Scientific Ltd., 618 Western Avenue, Park Royal, London W3 0TE, UK.
Hamilton Co., PO Box 10030, Reno, Nevada 89520-0012, USA.
  c/o Phase Separations Sales, Deeside Industrial Park, Deeside, Clwyd CH5 2NU, UK.
Hoefer Scientific Instruments, PO Box 77387-0387, 654 Minnesota Street, San Francisco, CA 94107, USA.
Hoefer UK Ltd., Newcastle, Staffs ST5 0TW, UK.
Kodak (Eastman Kodak Co.), PO Box 92822, LRPD-1001 Lee Road, Rochester, NY 14692-7073, USA.
  c/o Phase Separations Sales, Deeside Industrial Park, Deeside, Clwyd CH5 2NU, UK.
Merck AG, PO Box 4119, Frankfurterstrasse 250, D-6100 Darmstadt, Germany.
Merck Ltd., Merck House, Poole, Dorset BH15 1TD, UK.
Milligen (Millipore Intertech), PO Box 255, Bedford MA 01730, USA.
  (Millipore UK Ltd), The Boulevard, Blackmoor Lane, Watford, Herts WD1 8YW, UK.
M J Research Inc., 24 Bridge Street, Watertown, MA 02172, USA.
MWG-Biotech GmbH, Anzingerstrasse 7, D-8017 (85560) Ebersberg, Germany.
Nalgene (Nalge Co.), PO Box 20365, Rochester, NY 14602-0365, USA.
  c/o FSA Laboratory Supplies, Bishop Meadow Road, Loughborough, Leics. LE11 0RG, UK.
New England Biolabs (NBL), 32 Tozer Road, Beverley, MA 01915-5510, USA.
  c/o CP Labs Ltd, PO Box 22, Bishops Stortford, Herts CM23 3DH, UK.
New England Nuclear (NEN), Du Pont Co., NEN Research Products, 549 Albany Street, Boston, MA 02118, USA.
  c/o Du Pont UK Ltd, Wedgwood Way, Stevenage, Herts SG1 4QN, UK.
Nunc Inc., 2000 North Aurora Road, Naperville, IL 60563-1796, USA.
  c/o Life Technologies Ltd, Trident House, Renfrew Road, Paisley PA3 4EF, UK.

Appendix 1

Oncogene Science Inc., 106 Charles Lindbergh Boulevard, Uniondale, NY 11553, USA.
   c/o Cambridge Bioscience Ltd, 25 Signet Court, Stourbridge Common Business Centre, Swans Road, Cambridge CB5 8LA, UK.
Perkin Elmer-Cetus (The Perkin-Elmer Corporation), 761 Main Avenue, Norwalk, CT 0689-0251, USA.
Perkin-Elmer Ltd, Maxwell Road, Beaconsfield, Bucks HP9 1QA, UK.
Pharmacia Biosystems Ltd. (Biotechnology Division), Davy Avenue, Knowlhill, Milton Keynes MK5 8PH, UK.
Pharmacia LKB Biotechnology Inc., PO Box 1327, 800 Centennial Avenue, Piscataway, NJ 08855-1327, USA.
Pierce, 3747 North Meridan Road, PO Box 117, Rockford, IL 61105, USA.
   c/o Life Science Labs Ltd, Sedgewick Road, Luton, Beds LU4 9DT, UK.
Pierce Europe BV, PO Box 1512, 3260 BA Oud-Beijerland, The Netherlands.
PL Biochemicals; contact Pharmacia.
Promega Ltd, Delta House, Enterprise Road, Chilworth Research Centre, Southampton SO1 7NS, UK.
Promega, 2800 Woods Hollow Road, Madison, WI 53711-5399, USA.
Roth (Carl Roth GmbH), Schoemperlenstrasse 3, Postfach 211162, D-7500 Karlsruhe 21, Germany.
   c/o Techmate Ltd, 18 Bridgeturn Avenue, Old Wolverton, Milton Keynes MK12 5QL, UK.
Sandy Spring Instrument Co. Inc., Germantown, MD 21754, USA.
Sartorius AG, Postfach 3243, Weender Landstrasse 94-108, D-3400 Göttingen, Germany.
Sartorius Ltd, Longmead, Blenheim Road, Epsom, Surrey KT9 9QN, UK.
Sartorius Filters Inc., 30940 San Clemente Street, Hayward, CA 94544, USA.
Schleicher & Schuell, Postfach 4, D-3354 Dassell, Germany.
   c/o Anderman & Co. Ltd, 145 London Road, Kingston-upon-Thames, Surrey KT2 6NH, UK.
Seikagaku Corporation, Tokyo Yakugyo Building, 2-1-5 Nihonbashi-Honcho, Chuo-Ku, Tokyo, 103 Japan.
   c/o ICN Flow, Eagle House, Peregrine Business Park, Gomm Road, High Wycombe HP13 7DL, UK.
Serva Feinbiochemica GmbH & Co KG, PO Box 10 52 60, Carl Benzstrasse 7, D-6900 Heidelberg 1, Germany.
   c/o Universal Biologicals Ltd, 12–14 St Ann's Crescent, London SW18 2LS, UK.
Sigma Chemical Co. Ltd, Fancy Road, Poole, Dorset BH17 7NH, UK.
Sigma Inc., PO Box 14508, St Louis, MO 63178, USA.
Sorvall; contact Du Pont.
Spectrum Medical Industries Inc, 60916 Terminal Annex, Los Angeles, CA 90060, USA.
   c/o Orme Technology, PO Box 3, Stakehill Industrial Park, Middleton, Manchester M24 2RH, UK.

Appendix 1

Stratagene Inc., 11011 North Torrey Pines Road, La Jolla, CA 92037, USA.
Stratagene Ltd, Unit 140, Cambridge Innovation Centre, Milton Road, Cambridge CB4 4FG, UK.
Sylvania Lighting International, Otley Road, Charlestown, Shipley, West Yorkshire BD17 7SN, UK.
   c/o Osram Lighting, 100 Endicott Street, Danvers, MA 01923, USA.
Titertek; contact Flow (ICN).
United States Biochemical (USB) Corporation, PO Box 22400, Cleveland, OH 44122, USA.
   c/o Cambridge Bioscience, 25 Signet Court, Stourbridge Common Business Centre, Swans Road, Cambridge CB5 8LA, UK.
Waters (Millipore Intertech), PO Box 255, Bedford, MA 01730, USA.
Waters (Millipore UK Ltd), The Boulevard, Blackmoor Lane, Watford, Herts WD1 8YW, UK.
Whatman Scientific, Whatman House, St Leonards Road, Maidstone, Kent ME16 0LS, UK.
Whatman Scientific, 1000 North Tenth Street, Millville, NJ 08332, USA.
   c/o Jencons Scientific Ltd, Cherrycourt Way Industrial Estate, Stanbridge Road, Leighton Buzzard, Beds LU7 8UA, UK.

Contents of Volume II

1. 3' end-processing of mRNA
   Elmar Wahle and Walter Keller

2. Capping and methylation of mRNA
   Yasuhiro Furuichi and Aaron J. Shatkin

3. RNA editing in mitochondria
   Larry Simpson, Agda M. Simpson, and Beat Blum

4. Analysis of messenger RNA turnover in cell-free extracts from mammalian cells
   Jeff Ross

5. Ribosomal RNA processing in vertebrates
   Cathleen Enright and Barbara Sollner-Webb

6. Processing of transfer RNA precursors
   Chris L. Greer

7. Ribozymes
   David A. Shub, Craig L. Peebles, and Arnold Hampel

Index

affinity purification of
  poly(A)⁺ RNA 25-7, 61, 183-4
  snRNPs, snRNAs
    antisense oligonucleotides for 111-13
    biotin-streptavidin procedure 113-15, 122-3
    precleared extracts for 120-2
    recovery of RNA 116-17
    strategy 110-11
  snRNP-proteins 122-3
  splicing factors 89
  see also immunoaffinity purification of snRNPs
agarose (native) gel electrophoresis of
  DNA templates 3-5
  in vitro transcripts 14-16
  PCR products 7-8, 70
  RNA 14-16, 25
aniline, cleavage of modified RNA 138-9
antibodies
  for affinity purification of RNPs 142-9
  in analysis of RNA-protein interactions 159-63
  to cap structures 142, 145-6, 163
  in ELISA of snRNP 173
  immobilization on Protein A-Sepharose 142-5
  in immunoelectron microscopy of snRNPs 170-6
  in immunoprecipitation of snRNPs 159-63
  to snRNPs and snRNP proteins 145, 148-9, 163, 172, 175
antibody-Protein A Sepharose
  for immunoprecipitation of snRNPs 160-2
  preparation 142-4
antisense oligonucleotides
  in affinity selection of RNPs, snRNA 110-17, 120-3
  biotinylation 111-13
  chemical synthesis 104-10
  in depletion of splicing extracts 82, 117-20, 193-4
  for purification of snRNP proteins 122-3
  in targeted cleavage of RNA 82, 117-20, 123-6, 193-4

bacteriophage RNA polymerase 2-14, 131-3, 191
biotin-streptavidin affinity selection of RNP, RNA

analysis of RNP proteins 120-3
  factors determining 114
  procedure 113-15, 122-3
  recovering of RNA 116-17
  strategy 110-13
biotinylation of oligonucleotides 111-13
branch points in RNA mapping 49-53

calcium phosphate-DNA cotransfection 59
cap structures
  antibodies to 142, 145-7, 163
  incorporation into in vitro transcripts 11-12
  for protection of in vitro transcripts 11-12, 76
  in purification of splicing factors 89-91
  of snRNAs 141-2
cDNA
  amplification by PCR 41-2
  clones for splicing factors 98
  probes for nuclease S1 analysis 61-5
  synthesis 41
chemical modification-interference analysis of RNA
  cleavage of modified RNA 138-9
  end-labelling of RNA 135-6
  modification of RNA 136-8
  strategy 134-5
cleavage of RNA
  by aniline 138-9
  by RNase H 123-6, 193-4
cloning of splicing factors 98
complementation assay for splicing factors 86-8, 92-4
cross-linking, RNA-protein in RNP
  chemical 170-1
  UV 163-8

debranching, in vitro 51-3
depletion of extracts, using
  antibodies 82
  antisense oligonucleotides and RNase H 82, 117-20, 193-4
  genetic methods 194
diethylpyrocarbonate
  inactivation of RNase 2
  modification of RNA purine bases 134, 138

Index

DNA polymerase in
  cDNA synthesis 41
  3' end-labelling of DNA 33–4, 63
  PCR 41–2, 65–6

electron microscopy of snRNPs, *see* immuno-electron microscopy
electroblotting of RNA 130–1
electrophoresis, *see* agarose (native) gel electrophoresis, polyacrylamide–agarose (native) gel electrophoresis, polyacrylamide/urea (denaturing) gel electrophoresis, SDS-PAGE
ELISA of snRNPs, 173
end-labelling
  of DNA 33–5, 63–6
  of RNA 135–6
expression, of exogenous genes
  in mammalian cells
    analysis of RNA 61–73
    choice of cell line 58
    gene construct 60
    transfection procedure 59–60
    vectors 58–9
  in transgenic animals 60–1
  transient 58–60
  in yeast
    analysis of RNA 182–6
    transformation 180–1
    transplacement of mutant alleles 181–2
    vectors 179–80

FPLC
  of snRNPs 149–51
  of splicing factors 89, 93–4

genetic analysis of splicing 57, 179–94
gradient centrifugation
  CsCl
    of plasmid DNA 3–5
    of snRNP 88–9, 158–9
    of splicing factors 88–9, 92, 95–6
  glycerol
    of snRNPs 148–50
  sucrose
    of snRNP-antibody complexes 172–3

HeLa cell
  analysis of RNA splicing in 58
  debranching extract 51–2
  nuclear (splicing) extract 73–5, 84–6
  nucleoplasmic extract 92–3
  splicing factors, preparation 90–6
  transfection of 59

hybridization of probes
  in Northern blot analysis 48–9, 131–4
  in nuclease S1 analysis 36, 66–7
  in primer extension analysis 45–6, 184–5
  of riboprobes 133–4
  T_m, calculation of 37
hydrazine, modification of RNA pyrimidine bases 135, 137–8

immunoaffinity purification of snRNPs
  preparation of antibody–Sepharose matrix 142–5
  procedure 145–9
  strategy 142
  splicing factors 89–92
immunodepletion of splicing extracts 82
immunoelectron microscopy of snRNPs
  analysis 174–6
  formation of antibody–snRNP complexes 171–3
  sample preparation 174
  stabilisation of RNA–protein interactions 170–1
immunoprecipitation of snRNAs 159–63
interactive suppression in yeast 187–98
introns, mapping 5' ends 49–55
in vitro transcription, *see* transcription *in vitro*
ion exchange chromatography
  of snRNP proteins 156–8
  of snRNPs 149–51
  of splicing factors 88, 97

Klenow fragment (DNA polymerase I) for 3' end-labelling of DNA 33–4, 63

lariats, analysis of 50–5, 184–6

Northern blotting
  analysis of RNA 46–9, 130–4
  basic procedure 47–9, 130–4
  hybridization of probe 48–9, 131–4
  overview 46
  riboprobes for 132–3
  transfer of RNA 47–8, 130–1
nuclear extracts, *see* splicing extracts
nuclease S1 analysis
  basic procedure 35–7, 66–7
  end labelled probes for 33–5
  hybridization of probe 36, 66–7
  interpretation 37–8
  problems 37–8
  of RNA 32–8, 61–7

Index

strategy 32–3, 61–3
see also RNase H–nuclease S1 analysis
nucleic acids, removal from splicing extracts 88–9

oligo(2′-*O*-alkylribonucleotides), *see* antisense oligonucleotides
oligonucleotide
  affinity selection of RNPs 110–17, 120–3
  antisense 104–13
  biotinylation 111–13
  depletion of extracts 82, 117–20, 193–4
  end-labelling 34–45
  in primer extension analysis 44–6, 184–5
  probes 34–40, 48–9, 63–6, 111–13
  targeted cleavage of RNA 82, 117–20, 123–6, 193–4

PCR, *see* polymerase chain reaction, reverse transcription–polymerase chain reaction
plasmids
  for end-labelling probes 33–4
  for *in vitro* transcription 3–14, 132–3
  mammalian, expression 58–9
  preparation 3–5
  yeast, expression 179–82
polyacrylamide–agarose (native) gels
  for separation of snRNPs 128–30
polyacrylamide/urea (denaturing) gel electrophoresis
  of affinity-purified RNA 116–17
  of antisense oligonucleotides 108–10
  effect of polyacrylamide concentration on RNA mobility 50, 77–8
  of end-labelled probes 65–6
  of *in vitro* transcripts 17–21
  in Northern blots 47–8, 130–1
  in nuclease S1 analysis 37, 67
  in primer extension analysis 45–6, 184–6
  of proteins 87, 151–5, 162–3, 167
  of RNA 14–21, 37, 45–6, 76–81, 87, 126–8, 184–6
  of splicing factors 87, 91
  of splicing intermediates 18–19, 37, 45–6, 76–81, 87, 126–8, 184–6, 191
polymerase chain reaction (PCR)
  amplification of cDNA 41–2
  monitoring 70–1
  optimizing conditions 8, 70
  precautions 42–3
  preparation of DNA templates, probes 5–8, 65–6
  sequencing of products 71–3
  'touch-down' procedure 8, 70
polynucleotide kinase
  for 5′ end-labelling of DNA 34–5, 65

precursor RNA processing (PRP) genes, yeast 186–8
pre-mRNA splicing, *see* RNA splicing
primer extension analysis
  basic procedure 44–6, 184–5
  detection limits 46
  interpretation of data 186
  of intron ends in RNA 53–5
  of RNA branch points and lariats 50–3, 184–6
  of spliced RNA intermediates 43–6, 184–5
  overview of 43–4
probes
  antisense oligonucleotide 111–13
  biotinylated 111–13
  DNA 33–5, 48–9
  end-labelled 33–5, 63–6
  exon 46
  intron 46
  in Northern blotting 46–9, 131–4
  in nuclease S1 analysis 36–7, 61–7
  oligonucleotide 34–40, 48–9, 63–6, 111–13
  radiolabelled 33–5, 48–9, 63–6, 131–3
  riboprobes 131–3
  in RNase H–nuclease S1 analysis 38–40, 123–6, 193–4
  in targeted destruction of snRNA 38–40, 123–6, 193–4
protease digestion of snRNPs 167–8
Protein A–Sepharose, immobilisation of antibodies 142–4

recombinant splicing factors 95–8
reconstitution of snRNPs 156–9, 164–5, 192
reverse transcription
  for cDNA synthesis 41
  in primer extension 43–6, 184–6
  in RT-PCR 41–2, 70
reverse transcription–polymerase chain reaction (RT-PCR)
  in analysis of RNA splicing 40–3, 68–73
ribonucleoprotein (RNP) particles, *see* snRNPs
ribonucleotides, 2′-*O*-substituted, *see* antisense oligonucleotides
riboprobes
  in Northern blotting 133–4
  preparation 132–3
RNA
  analysis by
    combined RNase H and nuclease S1 38–40
    electrophoresis 14–21, 25, 126–8
    Northern blots 46–9, 130–4
    nuclease S1 32–8, 61–7
    primer extension 43–6, 50–5, 184–5
    RT-PCR 40–3, 68–73

207

Index

RNA (cont.)
  branch points, detection 49–53
  in cDNA synthesis 41
  chemical modification of 135–8
  cleavage
    chemical 138–9
    by RNase H 123–6, 193–4
  cross-linking to protein 163–8, 170–1
  electrophoresis
    denaturing 17–21, 126–8
    native 14–16, 25
  end-labelling 135–6
  hybridization 133–4
  lariats, analysis 50–5, 184–6
  ligase, end-labelling of RNA 135–6
  polymerase, bacteriophage 2–14, 17–18, 131–3, 191
  probes 131–3
  protection of 79–81, 134–9, 168–9
  purification
    using antisense oligonucleotides 116–17
    from cells, tissues 21–5, 61
    cytoplasmic 24–5
    using DNase 16–17
    by electrophoresis 14–19
    by immunoaffinity selection 152–3, 158–9
    in vitro transcripts 14–21
    poly(A)$^+$ RNA 25–7, 61, 183–4
    precautions against RNase 1–2, 31–2
    of snRNAs 27–8, 116–17, 152–3, 158–9
    from snRNPs 152–3, 158–9
    total 21–4, 182–3
    from yeast 182–4
  in RT-PCR 41–2, 70
  silver staining, in gels 155
  spliced, analysis of 32–49
  storage 31
  substrates, for *in vitro* splicing
    capped 11–12
    choice of 75–6
    heterogeneity of 17–18
    protection by capping 11–12, 76
    purification of 14–21
    quantification of 20–1
    radiolabelled 11–14, 132–3
    specific activity of 17, 20–1
    specific 3′ termini in 13–14
    stability 7, 11–12
    synthesis 8–14, 132–3
    unlabelled 10–11
  transcription, *see* transcription *in vitro*
RNA–protein interactions, analysis by
  chemical modification-interference 134–9
  immunoelectron microscopy 170–6
  immunoprecipitation 159–63
  RNase protection 168–9
  UV cross-linking 163–8

RNase
  digestion of cross-linked snRNPs 166–7
  inhibitors of 1–2, 87
  mapping of RNA–protein interactions 168–9
  precautions against 1–2, 31–2
  removal from solutions 1–2
RNase H
  analysis of RNA 38–40
  depletion of splicing extracts 82, 123–6, 193–4
  identification of spliced RNA 38–40, 79–81
  protection of RNA 79–81
  for specific 3′ termini in RNA transcripts 13–14
  targeted cleavage of RNA in RNPs 123–6, 193–4
RNase H–nuclease S1 analysis of RNA 38–40
RNase protection, in analysis of RNA–protein interactions 168–9
  splicing intermediates 79–81
RT-PCR, *see* reverse transcription–polymerase chain reaction
rTth DNA polymerase, in cDNA synthesis and PCR 41–2
run-off transcription, *see* transcription *in vitro*

SDS–PAGE of snRNP proteins 153–5, 162–3, 167
sedimentation coefficients of snRNPs 141, 146, 148–50
silver staining of RNA, protein 155
snRNA (small nuclear RNA)
  analysis 116–17, 126–8, 133–4, 151–5
  caps 89, 141–2
  depletion from splicing extracts 117–20, 193–4
  interactions with snRNP proteins 134–9, 159–76
  preparation 27–8, 110–17, 152–3, 158
  targeted cleavage in snRNPs 123–6, 193–4
snRNP proteins
  analysis 120–3, 151–8, 162, 167
  cross-linking to snRNA 163–8, 170–1
  interactions with snRNA 159–76
  preparation from snRNPs 122–3, 156–8
snRNPs (small nuclear ribonucleoprotein particles)
  analysis 128–30, 153–5, 162–3, 170–6
  cross-linking of RNA and protein in 163–8, 170–1
  depletion from splicing extracts 82, 117–20, 193–4
  ELISA 173

Index

isolation of
  proteins from 122–3
  snRNA from 152–3
preparation using
  antisense oligonucleotides 91, 110–17, 120–3
  FPLC 149–51
  gradient centrifugation 88–9, 148–50, 158–9, 172–3
  immunoaffinity selection 90–1, 142–9, 172–3
  immunoprecipitation 159–63
  ion exchange chromatography 149–51
protease digestion 165, 167–8
reconstitution of 156–9
RNA–protein interactions in 134–9, 159–76
RNase digestion 166–7
sedimentation coefficients 141, 146, 148–50
targeted cleavage of snRNA in 82, 117–20, 123–6, 193–4
see also U1, U2, U4, U5 and U6 snRNPs
SP6 phage polymerase, see bacteriophage RNA polymerase
spliceosomes, see snRNPs
splicing extracts
  affinity isolation of RNPs from 114–17, 120–3, 145–9
  biochemical fractionation 83–97
  choice of cells for 73–4
  depletion of RNPs and splicing factors from 82, 117–20, 123–6, 193–4
  importance of buffer composition 84, 88–9
  precleared 120–2
  preparation from
    HeLa cells 73–5, 84–6, 92–3
    yeast 190–4
  reconstitution of snRNPs in 156–9, 164–5, 193
  removal of nucleic acids from 88–9
  splicing factors, purification from 93–6
  targeted cleavage of RNA in 123–6, 193–4
splicing factors
  ASF 83, 92–5
  complementation assay for 86–8
  depletion of 82, 117–20, 193–4
  hnRNP AI 95–6
  HRF 97
  preparation of
    from cDNA clones 98
    from extracts 83–97
    importance of extract composition 84, 88–9
    problems in 83, 86–9
    of specific factors 90–7

strategy 83–9
recombinant 95–6, 98
SC35 83, 92–5
SF2 83, 92–5
SF5 95–6
SF33/4 83, 97
U2AF 83, 96–7
splicing of pre-mRNA
  analysis by
    complementation assay 86–8
    electrophoresis 76–81
    genetic approaches 57, 179–94
    interactive suppression 187–9
    Northern blotting 46–9, 130–4
    nuclease S1 33–8, 61–7
    primer extension 43–6, 50–5, 184–6
    RNase H-protection 79–81
    RNase H-nuclease S1 38–40
    RT-PCR 40–3, 68–73
  in extracts
    choice of substrate 75–6
    importance of extract composition 84, 88–9
    using microtitre plates 78–9
    preparation of extracts 73–5, 84–6, 92–3, 190–4
    substrates for 8–21, 132–3
    standard procedure 76–8
  in mammalian cells 58–60
  in transgenic animals 60–1
  in yeast 179–89
suppression, interactive, in yeast 187–9

T3, T7 phage polymerase, see bacteriophage RNA polymerase
Taq DNA polymerase, in PCR 65–6
T_m, calculation of 37
transcription *in vitro*
  using bacteriophage polymerases 8–14
  DNA template, preparation of 3–8
  efficiency of 14, 21
  generation of specific 3′ termini 13–14
  heterogeneity of transcripts 17–18
  incorporating cap analogue 11–12
  using non radioactive precursors 10–11
  premature termination 17–18
  procedure 10–14, 132–3
  purification of transcripts 14–21
  quantification 20–1
  using radioactive percursors 11–14, 132–3
transfection of mammalian cells 59–60
transformation of yeast 180–1, 187–9
transgenic animals, analysis of splicing 60–1
transient expression, analysis of splicing 58–60
transplacement of mutant alleles in yeast 181–2

Index

U1 snRNP 83, 88, 92, 117, 126, 141, 145–51, 155–6, 158–9, 162–3, 169–72, 174–6, 194
U2 snRNP 83, 117, 126, 141, 145–51, 156, 158, 193–4
U4 snRNP 83, 92, 117, 141, 145–51, 188
U5 snRNP 83, 92, 117, 120, 141, 145–51, 155–6, 188, 194
U6 snRNP 117, 141, 145–7, 150–1, 188, 193–4
UV cross-linking, RNA–protein
  extent of 165
  identification of proteins 165–7
  identification of RNA site 167–8
  method 164–5
  in snRNPs 163–8
  strategy 163–4

vectors for
  expression in yeast 179–82
  interactive suppression 187–9
  in vitro transcription 2–14, 131–3
  transient expression in mammalian cells 58–9
  transplacement of mutant alleles in yeast 181–2
  see also plasmids

yeast (*Saccharomyces cerevisiae*), in analysis of splicing
  expression vectors 179–80
  genetic analysis 186–9
  interactive suppression 187–9
  isolation of RNA 182–4
  primer extension 184–6
  splicing extracts 190–4
  transformation 180–1
  transplacement of mutant alleles 181–2

ORDER OTHER TITLES OF INTEREST TODAY

Price list for: UK, Europe, Rest of World (excluding US and Canada)

No.	Title	ISBN	Price
138.	**Plasmids (2/e)** Hardy, K.G. (Ed)		
	Spiralbound hardback	0-19-963445-9	£30.00
	Paperback	0-19-963444-0	£19.50
136.	**RNA Processing: Vol. II** Higgins, S.J. & Hames, B.D. (Eds)		
	Spiralbound hardback	0-19-963471-8	£30.00
	Paperback	0-19-963470-X	£19.50
135.	**RNA Processing: Vol. I** Higgins, S.J. & Hames, B.D. (Eds)		
	Spiralbound hardback	0-19-963344-4	£30.00
	Paperback	0-19-963343-6	£19.50
134.	**NMR of Macromolecules** Roberts, G.C.K. (Ed)		
	Spiralbound hardback	0-19-963225-1	£32.50
	Paperback	0-19-963224-3	£22.50
133.	**Gas Chromatography** Baugh, P. (Ed)		
	Spiralbound hardback	0-19-963272-3	£40.00
	Paperback	0-19-963271-5	£27.50
132.	**Essential Developmental Biology** Stern, C.D. & Holland, P.W.H. (Eds)		
	Spiralbound hardback	0-19-963423-8	£30.00
	Paperback	0-19-963422-X	£19.50
131.	**Cellular Interactions in Development** Hartley, D.A. (Ed)		
	Spiralbound hardback	0-19-963391-6	£30.00
	Paperback	0-19-963390-8	£18.50
129	**Behavioural Neuroscience: Volume II** Sahgal, A. (Ed)		
	Spiralbound hardback	0-19-963458-0	£32.50
	Paperback	0-19-963457-2	£22.50
128	**Behavioural Neuroscience: Volume I** Sahgal, A. (Ed)		
	Spiralbound hardback	0-19-963368-1	£32.50
	Paperback	0-19-963367-3	£22.50
127.	**Molecular Virology** Davison, A.J. & Elliott, R.M. (Eds)		
	Spiralbound hardback	0-19-963358-4	£35.00
	Paperback	0-19-963357-6	£25.00
126.	**Gene Targeting** Joyner, A.L. (Ed)		
	Spiralbound hardback	0-19-963407-6	£30.00
	Paperback	0-19-9634036-8	19.50
125.	**Glycobiology** Fukuda, M. & Kobata, A. (Eds)		
	Spiralbound hardback	0-19-963372-X	£32.50
	Paperback	0-19-963371-1	£22.50
124.	**Human Genetic Disease Analysis (2/e)** Davies, K.E. (Ed)		
	Spiralbound hardback	0-19-963309-6	£30.00
	Paperback	0-19-963308-8	£18.50
122.	**Immunocytochemistry** Beesley, J. (Ed)		
	Spiralbound hardback	0-19-963270-7	£35.00
	Paperback	0-19-963269-3	£22.50
123.	**Protein Phosphorylation** Hardie, D.G. (Ed)		
	Spiralbound hardback	0-19-963306-1	£32.50
	Paperback	0-19-963305-3	£22.50
121.	**Tumour Immunobiology** Gallagher, G., Rees, R.C. & others (Eds)		
	Spiralbound hardback	0-19-963370-3	£40.00
	Paperback	0-19-963369-X	£27.50
120.	**Transcription Factors** Latchman, D.S. (Ed)		
	Spiralbound hardback	0-19-963342-8	£30.00
	Paperback	0-19-963341-X	£19.50
119.	**Growth Factors** McKay, I. & Leigh, I. (Eds)		
	Spiralbound hardback	0-19-963360-6	£30.00
	Paperback	0-19-963359-2	£19.50
118.	**Histocompatibility Testing** Dyer, P. & Middleton, D. (Eds)		
	Spiralbound hardback	0-19-963364-9	£32.50
	Paperback	0-19-963363-0	£22.50
117.	**Gene Transcription** Hames, B.D. & Higgins, S.J. (Eds)		
	Spiralbound hardback	0-19-963292-8	£35.00
	Paperback	0-19-963291-X	£25.00
116.	**Electrophysiology** Wallis, D.I. (Ed)		
	Spiralbound hardback	0-19-963348-7	£32.50
	Paperback	0-19-963347-9	£22.50
115.	**Biological Data Analysis** Fry, J.C. (Ed)		
	Spiralbound hardback	0-19-963340-1	£50.00
	Paperback	0-19-963339-8	£27.50
114.	**Experimental Neuroanatomy** Bolam, J.P. (Ed)		
	Spiralbound hardback	0-19-963326-6	£32.50
	Paperback	0-19-963325-8	£22.50
113.	**Preparative Centrifugation** Rickwood, D. (Ed)		
	Spiralbound hardback	0-19-963208-1	£45.00
	Paperback	0-19-963211-1	£25.00
	Paperback	0-19-963099-2	£25.00
112.	**Lipid Analysis** Hamilton, R.J. & Hamilton, Shiela (Eds)		
	Spiralbound hardback	0-19-963098-4	£35.00
	Paperback	0-19-963099-2	£25.00
111.	**Haemopoiesis** Testa, N.G. & Molineux, G. (Eds)		
	Spiralbound hardback	0-19-963366-5	£32.50
	Paperback	0-19-963365-7	£22.50
110.	**Pollination Ecology** Dafni, A.		
	Spiralbound hardback	0-19-963299-5	£32.50
	Paperback	0-19-963298-7	£22.50
109.	**In Situ Hybridization** Wilkinson, D.G. (Ed)		
	Spiralbound hardback	0-19-963328-2	£30.00
	Paperback	0-19-963327-4	£18.50
108.	**Protein Engineering** Rees, A.R., Sternberg, M.J.E. & others (Eds)		
	Spiralbound hardback	0-19-963139-5	£35.00
	Paperback	0-19-963138-7	£25.00
107.	**Cell-Cell Interactions** Stevenson, B.R., Gallin, W.J. & others (Eds)		
	Spiralbound hardback	0-19-963319-3	£32.50
	Paperback	0-19-963318-5	£22.50
106.	**Diagnostic Molecular Pathology: Volume I** Herrington, C.S. & McGee, J. O'D. (Eds)		
	Spiralbound hardback	0-19-963237-5	£30.00
	Paperback	0-19-963236-7	£19.50
105.	**Biomechanics-Materials** Vincent, J.F.V. (Ed)		
	Spiralbound hardback	0-19-963223-5	£35.00
	Paperback	0-19-963222-7	£25.00
104.	**Animal Cell Culture (2/e)** Freshney, R.I. (Ed)		
	Spiralbound hardback	0-19-963212-X	£30.00
	Paperback	0-19-963213-8	£19.50
103.	**Molecular Plant Pathology: Volume II** Gurr, S.J., McPherson, M.J. & others (Eds)		
	Spiralbound hardback	0-19-963352-5	£32.50
	Paperback	0-19-963351-7	£22.50
102.	**Signal Transduction** Milligan, G. (Ed)		
	Spiralbound hardback	0-19-963296-0	£30.00
	Paperback	0-19-963295-2	£18.50
101.	**Protein Targeting** Magee, A.I. & Wileman, T. (Eds)		
	Spiralbound hardback	0-19-963206-5	£32.50
	Paperback	0-19-963210-3	£22.50
100.	**Diagnostic Molecular Pathology: Volume II: Cell and Tissue Genotyping** Herrington, C.S. & McGee, J.O'D. (Eds)		
	Spiralbound hardback	0-19-963239-1	£30.00
	Paperback	0-19-963238-3	£19.50
99.	**Neuronal Cell Lines** Wood, J.N. (Ed)		
	Spiralbound hardback	0-19-963346-0	£32.50
	Paperback	0-19-963345-2	£22.50

#	Title	Author/Editor	Format	ISBN	Price
98.	Neural Transplantation	Dunnett, S.B. & Björklund, A. (Eds)	Spiralbound hardback	0-19-963286-3	£30.00
			Paperback	0-19-963285-5	£19.50
97.	Human Cytogenetics: Volume II: Malignancy and Acquired Abnormalities (2/e)	Rooney, D.E. & Czepulkowski, B.H. (Eds)	Spiralbound hardback	0-19-963290-1	£30.00
			Paperback	0-19-963289-8	£22.50
96.	Human Cytogenetics: Volume I: Constitutional Analysis (2/e)	Rooney, D.E. & Czepulkowski, B.H. (Eds)	Spiralbound hardback	0-19-963288-X	£30.00
			Paperback	0-19-963287-1	£22.50
95.	Lipid Modification of Proteins	Hooper, N.M. & Turner, A.J. (Eds)	Spiralbound hardback	0-19-963274-X	£32.50
			Paperback	0-19-963273-1	£22.50
94.	Biomechanics-Structures and Systems	Biewener, A.A. (Ed)	Spiralbound hardback	0-19-963268-5	£42.50
			Paperback	0-19-963267-7	£25.00
93.	Lipoprotein Analysis	Converse, C.A. & Skinner, E.R. (Eds)	Spiralbound hardback	0-19-963192-1	£30.00
			Paperback	0-19-963231-6	£19.50
92.	Receptor-Ligand Interactions	Hulme, E.C. (Ed)	Spiralbound hardback	0-19-963090-9	£35.00
			Paperback	0-19-963091-7	£27.50
91.	Molecular Genetic Analysis of Populations	Hoelzel, A.R. (Ed)	Spiralbound hardback	0-19-963278-2	£32.50
			Paperback	0-19-963277-4	£22.50
90.	Enzyme Assays	Eisenthal, R. & Danson, M.J. (Eds)	Spiralbound hardback	0-19-963142-5	£35.00
			Paperback	0-19-963143-3	£25.00
89.	Microcomputers in Biochemistry	Bryce, C.F.A. (Ed)	Spiralbound hardback	0-19-963253-7	£30.00
			Paperback	0-19-963252-9	£19.50
88.	The Cytoskeleton	Carraway, K.L. & Carraway, C.A.C. (Eds)	Spiralbound hardback	0-19-963257-X	£30.00
			Paperback	0-19-963256-1	£19.50
87.	Monitoring Neuronal Activity	Stamford, J.A. (Ed)	Spiralbound hardback	0-19-963244-8	£30.00
			Paperback	0-19-963243-X	£19.50
86.	Crystallization of Nucleic Acids and Proteins	Ducruix, A. & Giegé, R. (Eds)	Spiralbound hardback	0-19-963245-6	£35.00
			Paperback	0-19-963246-4	£25.00
85.	Molecular Plant Pathology: Volume I	Gurr, S.J., McPherson, M.J. & others (Eds)	Spiralbound hardback	0-19-963103-4	£30.00
			Paperback	0-19-963102-6	£19.50
84.	Anaerobic Microbiology	Levett, P.N. (Ed)	Spiralbound hardback	0-19-963204-9	£32.50
			Paperback	0-19-963262-6	£22.50
83.	Oligonucleotides and Analogues	Eckstein, F. (Ed)	Spiralbound hardback	0-19-963280-4	£32.50
			Paperback	0-19-963279-0	£22.50
82.	Electron Microscopy in Biology	Harris, R. (Ed)	Spiralbound hardback	0-19-963219-7	£32.50
			Paperback	0-19-963215-4	£22.50
81.	Essential Molecular Biology: Volume II	Brown, T.A. (Ed)	Spiralbound hardback	0-19-963112-3	£32.50
			Paperback	0-19-963113-1	£22.50
80.	Cellular Calcium	McCormack, J.G. & Cobbold, P.H. (Eds)	Spiralbound hardback	0-19-963131-X	£35.00
			Paperback	0-19-963130-1	£25.00
79.	Protein Architecture	Lesk, A.M.	Spiralbound hardback	0-19-963054-2	£32.50
			Paperback	0-19-963055-0	£22.50
78.	Cellular Neurobiology	Chad, J. & Wheal, H. (Eds)	Spiralbound hardback	0-19-963106-9	£32.50
			Paperback	0-19-963107-7	£22.50
77.	PCR	McPherson, M.J., Quirke, P. & others (Eds)	Spiralbound hardback	0-19-963226-X	£30.00
			Paperback	0-19-963196-4	£19.50
76.	Mammalian Cell Biotechnology	Butler, M. (Ed)	Spiralbound hardback	0-19-963207-3	£30.00
			Paperback	0-19-963209-X	£19.50
75.	Cytokines	Balkwill, F.R. (Ed)	Spiralbound hardback	0-19-963218-9	£35.00
			Paperback	0-19-963214-6	£25.00
74.	Molecular Neurobiology	Chad, J. & Wheal, H. (Eds)	Spiralbound hardback	0-19-963108-5	£30.00
			Paperback	0-19-963109-3	£19.50
73.	Directed Mutagenesis	McPherson, M.J. (Ed)	Spiralbound hardback	0-19-963141-7	£30.00
			Paperback	0-19-963140-9	£19.50
72.	Essential Molecular Biology: Volume I	Brown, T.A. (Ed)	Spiralbound hardback	0-19-963110-7	£32.50
			Paperback	0-19-963111-5	£22.50
71.	Peptide Hormone Action	Siddle, K. & Hutton, J.C.	Spiralbound hardback	0-19-963070-4	£32.50
			Paperback	0-19-963071-2	£22.50
70.	Peptide Hormone Secretion	Hutton, J.C. & Siddle, K. (Eds)	Spiralbound hardback	0-19-963068-2	£35.00
			Paperback	0-19-963069-0	£25.00
69.	Postimplantation Mammalian Embryos	Copp, A.J. & Cockroft, D.L. (Eds)	Spiralbound hardback	0-19-963088-7	£15.00
			Paperback	0-19-963089-5	£12.50
68.	Receptor-Effector Coupling	Hulme, E.C. (Ed)	Spiralbound hardback	0-19-963094-1	£30.00
			Paperback	0-19-963095-X	£19.50
67.	Gel Electrophoresis of Proteins (2/e)	Hames, B.D. & Rickwood, D. (Eds)	Spiralbound hardback	0-19-963074-7	£35.00
			Paperback	0-19-963075-5	£25.00
66.	Clinical Immunology	Gooi, H.C. & Chapel, H. (Eds)	Spiralbound hardback	0-19-963086-0	£32.50
			Paperback	0-19-963087-9	£22.50
65.	Receptor Biochemistry	Hulme, E.C. (Ed)	Paperback	0-19-963093-3	£25.00
64.	Gel Electrophoresis of Nucleic Acids (2/e)	Rickwood, D. & Hames, B.D. (Eds)	Spiralbound hardback	0-19-963082-8	£32.50
			Paperback	0-19-963083-6	£22.50
63.	Animal Virus Pathogenesis	Oldstone, M.B.A. (Ed)	Spiralbound hardback	0-19-963100-X	£15.00
			Paperback	0-19-963101-8	£12.50
62.	Flow Cytometry	Ormerod, M.G. (Ed)	Paperback	0-19-963053-4	£22.50
61.	Radioisotopes in Biology	Slater, R.J. (Ed)	Spiralbound hardback	0-19-963080-1	£32.50
			Paperback	0-19-963081-X	£22.50
60.	Biosensors	Cass, A.E.G. (Ed)	Spiralbound hardback	0-19-963046-1	£30.00
			Paperback	0-19-963047-X	£19.50
59.	Ribosomes and Protein Synthesis	Spedding, G. (Ed)	Spiralbound hardback	0-19-963104-2	£15.00
			Paperback	0-19-963105-0	£12.50
58.	Liposomes	New, R.R.C. (Ed)	Spiralbound hardback	0-19-963076-3	£35.00
			Paperback	0-19-963077-1	£22.50
57.	Fermentation	McNeil, B. & Harvey, L.M. (Eds)	Spiralbound hardback	0-19-963044-5	£30.00
			Paperback	0-19-963045-3	£19.50
56.	Protein Purification Applications	Harris, E.L.V. & Angal, S. (Eds)	Spiralbound hardback	0-19-963022-4	£30.00
			Paperback	0-19-963023-2	£18.50
55.	Nucleic Acids Sequencing	Howe, C.J. & Ward, E.S. (Eds)	Spiralbound hardback	0-19-963056-9	£30.00
			Paperback	0-19-963057-7	£19.50
54.	Protein Purification Methods	Harris, E.L.V. & Angal, S. (Eds)	Spiralbound hardback	0-19-963002-X	£30.00
			Paperback	0-19-963003-8	£19.50
53.	Solid Phase Peptide Synthesis	Atherton, E. & Sheppard, R.C.	Spiralbound hardback	0-19-963066-6	£15.00
			Paperback	0-19-963067-4	£12.50
52.	Medical Bacteriology	Hawkey, P.M. & Lewis, D.A. (Eds)	Paperback	0-19-963009-7	£25.00
51.	Proteolytic Enzymes	Beynon, R.J. & Bond, J.S. (Eds)	Spiralbound hardback	0-19-963058-5	£30.00
			Paperback	0-19-963059-3	£19.50
50.	Medical Mycology	Evans, E.G.V. & Richardson, M.D. (Eds)	Spiralbound hardback	0-19-963010-0	£37.50
			Paperback	0-19-963011-9	£25.00
49.	Computers in Microbiology	Bryant, T.N. & Wimpenny, J.W.T. (Eds)	Paperback	0-19-963015-1	£12.50

No.	Title	Editor(s)	Format	ISBN	Price
48.	Protein Sequencing	Findlay, J.B.C. & Geisow, M.J. (Eds)	Spiralbound hardback	0-19-963012-7	£15.00
			Paperback	0-19-963013-5	£12.50
47.	Cell Growth and Division	Baserga, R. (Ed)	Spiralbound hardback	0-19-963026-7	£15.00
			Paperback	0-19-963027-5	£12.50
46.	Protein Function	Creighton, T.E. (Ed)	Spiralbound hardback	0-19-963006-2	£32.50
			Paperback	0-19-963007-0	£22.50
45.	Protein Structure	Creighton, T.E. (Ed)	Spiralbound hardback	0-19-963000-3	£32.50
			Paperback	0-19-963001-1	£22.50
44.	Antibodies: Volume II	Catty, D. (Ed)	Spiralbound hardback	0-19-963018-6	£30.00
			Paperback	0-19-963019-4	£19.50
43.	HPLC of Macromolecules	Oliver, R.W.A. (Ed)	Spiralbound hardback	0-19-963020-8	£30.00
			Paperback	0-19-963021-6	£19.50
42.	Light Microscopy in Biology	Lacey, A.J. (Ed)	Spiralbound hardback	0-19-963036-4	£30.00
			Paperback	0-19-963037-2	£19.50
41.	Plant Molecular Biology	Shaw, C.H. (Ed)	Paperback	1-85221-056-7	£12.50
40.	Microcomputers in Physiology	Fraser, P.J. (Ed)	Spiralbound hardback	1-85221-129-6	£15.00
			Paperback	1-85221-130-X	£12.50
39.	Genome Analysis	Davies, K.E. (Ed)	Spiralbound hardback	1-85221-109-1	£30.00
			Paperback	1-85221-110-5	£18.50
38.	Antibodies: Volume I	Catty, D. (Ed)	Paperback	0-947946-85-3	£19.50
37.	Yeast	Campbell, I. & Duffus, J.H. (Eds)	Paperback	0-947946-79-9	£12.50
36.	Mammalian Development	Monk, M. (Ed)	Hardback	1-85221-030-3	£15.00
			Paperback	1-85221-029-X	£12.50
35.	Lymphocytes	Klaus, G.G.B. (Ed)	Hardback	1-85221-018-4	£30.00
34.	Lymphokines and Interferons	Clemens, M.J., Morris, A.G. & others (Eds)	Paperback	1-85221-035-4	£12.50
33.	Mitochondria	Darley-Usmar, V.M., Rickwood, D. & others (Eds)	Hardback	1-85221-034-6	£32.50
			Paperback	1-85221-033-8	£22.50
32.	Prostaglandins and Related Substances	Benedetto, C., McDonald-Gibson, R.G. & others (Eds)	Hardback	1-85221-032-X	£15.00
			Paperback	1-85221-031-1	£12.50
31.	DNA Cloning: Volume III	Glover, D.M. (Ed)	Hardback	1-85221-049-4	£15.00
			Paperback	1-85221-048-6	£12.50
30.	Steroid Hormones	Green, B. & Leake, R.E. (Eds)	Paperback	0-947946-53-5	£19.50
29.	Neurochemistry	Turner, A.J. & Bachelard, H.S. (Eds)	Hardback	1-85221-028-1	£15.00
			Paperback	1-85221-027-3	£12.50
28.	Biological Membranes	Findlay, J.B.C. & Evans, W.H. (Eds)	Hardback	0-947946-84-5	£15.00
			Paperback	0-947946-83-7	£12.50
27.	Nucleic Acid and Protein Sequence Analysis	Bishop, M.J. & Rawlings, C.J. (Eds)	Hardback	1-85221-007-9	£35.00
			Paperback	1-85221-006-0	£25.00
26.	Electron Microscopy in Molecular Biology	Sommerville, J. & Scheer, U. (Eds)	Hardback	0-947946-64-0	£15.00
			Paperback	0-947946-54-3	£12.50
25.	Teratocarcinomas and Embryonic Stem Cells	Robertson, E.J. (Ed)	Paperback	1-85221-004-4	£19.50
24.	Spectrophotometry and Spectrofluorimetry	Harris, D.A. & Bashford, C.L. (Eds)	Hardback	0-947946-69-1	£15.00
			Paperback	0-947946-46-2	£12.50
23.	Plasmids	Hardy, K.G. (Ed)	Paperback	0-947946-81-0	£12.50
22.	Biochemical Toxicology	Snell, K. & Mullock, B. (Eds)	Paperback	0-947946-52-7	£12.50
19.	Drosophila	Roberts, D.B. (Ed)	Hardback	0-947946-66-7	£32.50
			Paperback	0-947946-45-4	£22.50
17.	Photosynthesis: Energy Transduction	Hipkins, M.F. & Baker, N.R. (Eds)	Hardback	0-947946-63-2	£15.00
			Paperback	0-947946-51-9	£12.50
16.	Human Genetic Diseases	Davies, K.E. (Ed)	Hardback	0-947946-76-4	£15.00
			Paperback	0-947946-75-6	£12.50
14.	Nucleic Acid Hybridisation	Hames, B.D. & Higgins, S.J. (Eds)	Hardback	0-947946-61-6	£15.00
			Paperback	0-947946-23-3	£12.50
13.	Immobilised Cells and Enzymes	Woodward, J. (Ed)	Hardback	0-947946-60-8	£15.00
12.	Plant Cell Culture	Dixon, R.A. (Ed)	Paperback	0-947946-22-5	£19.50
11a.	DNA Cloning: Volume I	Glover, D.M. (Ed)	Paperback	0-947946-18-7	£12.50
11b.	DNA Cloning: Volume II	Glover, D.M. (Ed)	Paperback	0-947946-19-5	£12.50
10.	Virology	Mahy, B.W.J. (Ed)	Paperback	0-904147-78-9	£19.50
9.	Affinity Chromatography	Dean, P.D.G., Johnson, W.S. & others (Eds)	Paperback	0-904147-71-1	£19.50
7.	Microcomputers in Biology	Ireland, C.R. & Long, S.P. (Eds)	Paperback	0-904147-57-6	£18.00
6.	Oligonucleotide Synthesis	Gait, M.J. (Ed)	Paperback	0-904147-74-6	£18.50
5.	Transcription and Translation	Hames, B.D. & Higgins, S.J. (Eds)	Paperback	0-904147-52-5	£12.50
3.	Iodinated Density Gradient Media	Rickwood, D. (Ed)	Paperback	0-904147-51-7	£12.50

Sets

Title	Editor(s)	Format	ISBN	Price
Essential Molecular Biology: 2 vol set	Brown, T.A. (Ed)	Spiralbound hardback	0-19-963114-X	£58.00
		Paperback	0-19-963115-8	£40.00
Antibodies: 2 vol set	Catty, D. (Ed)	Paperback	0-19-963063-1	£33.00
Cellular and Molecular Neurobiology: 2 vol set	Chad, J. & Wheal, H. (Eds)	Spiralbound hardback	0-19-963255-3	£56.00
		Paperback	0-19-963256-9	£38.00
Protein Structure and Protein Function: 2 vol set	Creighton, T.E. (Ed)	Spiralbound hardback	0-19-963064-X	£55.00
		Paperback	0-19-963065-8	£38.00
DNA Cloning: 2 vol set	Glover, D.M. (Ed)	Paperback	1-85221-069-9	£30.00
Molecular Plant Pathology: 2 vol set	Gurr, S.J., McPherson, M.J. & others (Eds)	Spiralbound hardback	0-19-963354-1	£56.00
		Paperback	0-19-963353-3	£37.00
Protein Purification Methods, and Protein Purification Applications: 2 vol set	Harris, E.L.V. & Angal, S. (Eds)	Spiralbound hardback	0-19-963048-8	£48.00
		Paperback	0-19-963049-6	£32.00
Diagnostic Molecular Pathology: 2 vol set	Herrington, C.S. & McGee, J. O'D. (Eds)	Spiralbound hardback	0-19-963241-3	£54.00
		Paperback	0-19-963240-5	£35.00
RNA Processing: 2 vol set	Higgins, S.J. & Hames, B.D. (Eds)	Spiralbound hardback	0-19-963473-4	£54.00
		Paperback	0-19-963472-6	£35.00
Receptor Biochemistry; Receptor-Effector Coupling; Receptor-Ligand Interactions: 3 vol set	Hulme, E.C. (Ed)	Paperback	0-19-963097-6	£62.50
Human Cytogenetics: 2 vol set (2/e)	Rooney, D.E. & Czepulkowski, B.H. (Eds)	Hardback	0-19-963314-2	£58.50
		Paperback	0-19-963313-4	£40.50
Behavioural Neuroscience: 2 vol set	Sahgal, A. (Ed)	Spiralbound hardback	0-19-963460-2	£58.00
		Paperback	0-19-963459-9	£40.00
Peptide Hormone Secretion/Peptide Hormone Action: 2 vol set	Siddle, K. & Hutton, J.C. (Eds)	Spiralbound hardback	0-19-963072-0	£55.00
		Paperback	0-19-963073-9	£38.00

ORDER OTHER TITLES OF INTEREST TODAY

Price list for: USA and Canada

128.	**Behavioural Neuroscience: Volume I** Sahgal, A. (Ed)			
......	Spiralbound hardback	0-19-963368-1	$57.00	
......	Paperback	0-19-963367-3	$37.00	
127.	**Molecular Virology** Davison, A.J. & Elliott, R.M. (Eds)			
......	Spiralbound hardback	0-19-963358-4	$49.00	
......	Paperback	0-19-963357-6	$32.00	
126.	**Gene Targeting** Joyner, A.L. (Ed)			
......	Spiralbound hardback	0-19-963407-6	$49.00	
......	Paperback	0-19-9634036-8	$34.00	
124.	**Human Genetic Disease Analysis (2/e)** Davies, K.E. (Ed)			
......	Spiralbound hardback	0-19-963309-6	$54.00	
......	Paperback	0-19-963308-8	$33.00	
123.	**Protein Phosphorylation** Hardie, D.G. (Ed)			
......	Spiralbound hardback	0-19-963306-1	$65.00	
......	Paperback	0-19-963305-3	$45.00	
122.	**Immunocytochemistry** Beesley, J. (Ed)			
......	Spiralbound hardback	0-19-963270-7	$62.00	
......	Paperback	0-19-963269-3	$42.00	
121.	**Tumour Immunobiology** Gallagher, G., Rees, R.C. & others (Eds)			
......	Spiralbound hardback	0-19-963370-3	$72.00	
......	Paperback	0-19-963369-X	$50.00	
120.	**Transcription Factors** Latchman, D.S. (Ed)			
......	Spiralbound hardback	0-19-963342-8	$48.00	
......	Paperback	0-19-963341-X	$31.00	
119.	**Growth Factors** McKay, I. & Leigh, I. (Eds)			
......	Spiralbound hardback	0-19-963360-6	$48.00	
......	Paperback	0-19-963359-2	$31.00	
118.	**Histocompatibility Testing** Dyer, P. & Middleton, D. (Eds)			
......	Spiralbound hardback	0-19-963364-9	$60.00	
......	Paperback	0-19-963363-0	$41.00	
117.	**Gene Transcription** Hames, B.D. & Higgins, S.J. (Eds)			
......	Spiralbound hardback	0-19-963292-8	$72.00	
......	Paperback	0-19-963291-X	$50.00	
116.	**Electrophysiology** Wallis, D.I. (Ed)			
......	Spiralbound hardback	0-19-963348-7	$56.00	
......	Paperback	0-19-963347-9	$39.00	
115.	**Biological Data Analysis** Fry, J.C. (Ed)			
......	Spiralbound hardback	0-19-963340-1	$80.00	
......	Paperback	0-19-963339-8	$60.00	
114.	**Experimental Neuroanatomy** Bolam, J.P. (Ed)			
......	Spiralbound hardback	0-19-963326-6	$59.00	
......	Paperback	0-19-963325-8	$39.00	
113.	**Preparative Centrifugation** Rickwood, D. (Ed)			
......	Spiralbound hardback	0-19-963208-1	$78.00	
......	Paperback	0-19-963211-1	$44.00	
111.	**Haemopoiesis** Testa, N.G. & Molineux, G. (Eds)			
......	Spiralbound hardback	0-19-963366-5	$59.00	
......	Paperback	0-19-963365-7	$39.00	
110.	**Pollination Ecology** Dafni, A.			
......	Spiralbound hardback	0-19-963299-5	$56.95	
......	Paperback	0-19-963298-7	$39.95	
109.	**In Situ Hybridization** Wilkinson, D.G. (Ed)			
......	Spiralbound hardback	0-19-963328-2	$58.00	
......	Paperback	0-19-963327-4	$36.00	
108.	**Protein Engineering** Rees, A.R., Sternberg, M.J.E. & others (Eds)			
......	Spiralbound hardback	0-19-963139-5	$64.00	
......	Paperback	0-19-963138-7	$44.00	
107.	**Cell-Cell Interactions** Stevenson, B.R., Gallin, W.J. & others (Eds)			
......	Spiralbound hardback	0-19-963319-3	$55.00	
......	Paperback	0-19-963318-5	$38.00	
106.	**Diagnostic Molecular Pathology: Volume I** Herrington, C.S. & McGee, J. O'D. (Eds)			
......	Spiralbound hardback	0-19-963237-5	$50.00	
......	Paperback	0-19-963236-7	$33.00	
105.	**Biomechanics-Materials** Vincent, J.F.V. (Ed)			
......	Spiralbound hardback	0-19-963223-5	$70.00	
......	Paperback	0-19-963222-7	$50.00	
104.	**Animal Cell Culture (2/e)** Freshney, R.I. (Ed)			
......	Spiralbound hardback	0-19-963212-X	$55.00	
......	Paperback	0-19-963213-8	$35.00	
103.	**Molecular Plant Pathology: Volume II** Gurr, S.J., McPherson, M.J. & others (Eds)			
......	Spiralbound hardback	0-19-963352-5	$65.00	
......	Paperback	0-19-963351-7	$45.00	
102.	**Signal Transduction** Milligan, G. (Ed)			
......	Spiralbound hardback	0-19-963296-0	$60.00	
......	Paperback	0-19-963295-2	$38.00	
101.	**Protein Targeting** Magee, A.I. & Wileman, T. (Eds)			
......	Spiralbound hardback	0-19-963206-5	$75.00	
......	Paperback	0-19-963210-3	$50.00	
100.	**Diagnostic Molecular Pathology: Volume II: Cell and Tissue Genotyping** Herrington, C.S. & McGee, J.O'D. (Eds)			
......	Spiralbound hardback	0-19-963239-1	$60.00	
......	Paperback	0-19-963238-3	$39.00	
99.	**Neuronal Cell Lines** Wood, J.N. (Ed)			
......	Spiralbound hardback	0-19-963346-0	$68.00	
......	Paperback	0-19-963345-2	$48.00	
98.	**Neural Transplantation** Dunnett, S.B. & Björklund, A. (Eds)			
......	Spiralbound hardback	0-19-963286-3	$69.00	
......	Paperback	0-19-963285-5	$42.00	
97.	**Human Cytogenetics: Volume II: Malignancy and Acquired Abnormalities (2/e)** Rooney, D.E. & Czepulkowski, B.H. (Eds)			
......	Spiralbound hardback	0-19-963290-1	$75.00	
......	Paperback	0-19-963289-8	$50.00	
96.	**Human Cytogenetics: Volume I: Constitutional Analysis (2/e)** Rooney, D.E. & Czepulkowski, B.H. (Eds)			
......	Spiralbound hardback	0-19-963288-X	$75.00	
......	Paperback	0-19-963287-1	$50.00	
95.	**Lipid Modification of Proteins** Hooper, N.M. & Turner, A.J. (Eds)			
......	Spiralbound hardback	0-19-963274-X	$75.00	
......	Paperback	0-19-963273-1	$50.00	
94.	**Biomechanics-Structures and Systems** Biewener, A.A. (Ed)			
......	Spiralbound hardback	0-19-963268-5	$85.00	
......	Paperback	0-19-963267-7	$50.00	
93.	**Lipoprotein Analysis** Converse, C.A. & Skinner, E.R. (Eds)			
......	Spiralbound hardback	0-19-963192-1	$65.00	
......	Paperback	0-19-963231-6	$42.00	
92.	**Receptor-Ligand Interactions** Hulme, E.C. (Ed)			
......	Spiralbound hardback	0-19-963090-9	$75.00	
......	Paperback	0-19-963091-7	$50.00	
91.	**Molecular Genetic Analysis of Populations** Hoelzel, A.R. (Ed)			
......	Spiralbound hardback	0-19-963278-2	$65.00	
......	Paperback	0-19-963277-4	$45.00	

ORDER FORM for UK, Europe and Rest of World

(Excluding USA and Canada)

Qty	ISBN	Author	Title	Amount
			P&P	
			*VAT	
			TOTAL	

Please add postage and packing: £1.75 for UK orders under £20; £2.75 for UK orders over £20; overseas orders add 10% of total.

* EC customers please note that VAT must be added (excludes UK customers)

Name ....

Address ....

....

.... Post code ....

[ ] Please charge £ .... to my credit card

Access/VISA/Eurocard/AMEX/Diners Club (circle appropriate card)

Card No .... Expiry date ....

Signature ....

Credit card account address if different from above:

....

.... Postcode ....

[ ] I enclose a cheque for £.......................

Please return this form to: OUP Distribution Services, Saxon Way West, Corby, Northants NN18 9ES, UK

OR ORDER BY CREDIT CARD HOTLINE: Tel +44-(0)536-741519 or Fax +44-(0)536-746337

#	Title	Editor(s)/Author(s)	Format	ISBN	Price
90.	Enzyme Assays	Eisenthal, R. & Danson, M.J. (Eds)	Spiralbound hardback	0-19-963142-5	$68.00
			Paperback	0-19-963143-3	$48.00
89.	Microcomputers in Biochemistry	Bryce, C.F.A. (Ed)	Spiralbound hardback	0-19-963253-7	$60.00
			Paperback	0-19-963252-9	$40.00
88.	The Cytoskeleton	Carraway, K.L. & Carraway, C.A.C. (Eds)	Spiralbound hardback	0-19-963257-X	$60.00
			Paperback	0-19-963256-1	$40.00
87.	Monitoring Neuronal Activity	Stamford, J.A. (Ed)	Spiralbound hardback	0-19-963244-8	$60.00
			Paperback	0-19-963243-X	$40.00
86.	Crystallization of Nucleic Acids and Proteins	Ducruix, A. & Giegé, R. (Eds)	Spiralbound hardback	0-19-963245-6	$60.00
			Paperback	0-19-963246-4	$50.00
85.	Molecular Plant Pathology: Volume I	Gurr, S.J., McPherson, M.J. & others (Eds)	Spiralbound hardback	0-19-963103-4	$60.00
			Paperback	0-19-963102-6	$40.00
84.	Anaerobic Microbiology	Levett, P.N. (Ed)	Spiralbound hardback	0-19-963204-9	$75.00
			Paperback	0-19-963262-6	$45.00
83.	Oligonucleotides and Analogues	Eckstein, F. (Ed)	Spiralbound hardback	0-19-963280-4	$65.00
			Paperback	0-19-963279-0	$45.00
82.	Electron Microscopy in Biology	Harris, R. (Ed)	Spiralbound hardback	0-19-963219-7	$65.00
			Paperback	0-19-963215-4	$45.00
81.	Essential Molecular Biology: Volume II	Brown, T.A. (Ed)	Spiralbound hardback	0-19-963112-3	$65.00
			Paperback	0-19-963113-1	$45.00
80.	Cellular Calcium	McCormack, J.G. & Cobbold, P.H. (Eds)	Spiralbound hardback	0-19-963131-X	$75.00
			Paperback	0-19-963130-1	$50.00
79.	Protein Architecture	Lesk, A.M.	Spiralbound hardback	0-19-963054-2	$65.00
			Paperback	0-19-963055-0	$45.00
78.	Cellular Neurobiology	Chad, J. & Wheal, H. (Eds)	Spiralbound hardback	0-19-963106-9	$73.00
			Paperback	0-19-963107-7	$43.00
77.	PCR	McPherson, M.J., Quirke, P. & others (Eds)	Spiralbound hardback	0-19-963226-X	$55.00
			Paperback	0-19-963196-4	$40.00
76.	Mammalian Cell Biotechnology	Butler, M. (Ed)	Spiralbound hardback	0-19-963207-3	$60.00
			Paperback	0-19-963209-X	$40.00
75.	Cytokines	Balkwill, F.R. (Ed)	Spiralbound hardback	0-19-963218-9	$64.00
			Paperback	0-19-963214-6	$44.00
74.	Molecular Neurobiology	Chad, J. & Wheal, H. (Eds)	Spiralbound hardback	0-19-963108-5	$56.00
			Paperback	0-19-963109-3	$36.00
73.	Directed Mutagenesis	McPherson, M.J. (Ed)	Spiralbound hardback	0-19-963141-7	$55.00
			Paperback	0-19-963140-9	$35.00
72.	Essential Molecular Biology: Volume I	Brown, T.A. (Ed)	Spiralbound hardback	0-19-963110-7	$65.00
			Paperback	0-19-963111-5	$45.00
71.	Peptide Hormone Action	Siddle, K. & Hutton, J.C.	Spiralbound hardback	0-19-963070-4	$70.00
			Paperback	0-19-963071-2	$50.00
70.	Peptide Hormone Secretion	Hutton, J.C. & Siddle, K. (Eds)	Spiralbound hardback	0-19-963068-2	$70.00
			Paperback	0-19-963069-0	$50.00
69.	Postimplantation Mammalian Embryos	Copp, A.J. & Cockroft, D.L. (Eds)	Spiralbound hardback	0-19-963088-7	$70.00
			Paperback	0-19-963089-5	$50.00
68.	Receptor-Effector Coupling	Hulme, E.C. (Ed)	Spiralbound hardback	0-19-963094-1	$70.00
			Paperback	0-19-963095-X	$50.00
67.	Gel Electrophoresis of Proteins (2/e)	Hames, B.D. & Rickwood, D. (Eds)	Spiralbound hardback	0-19-963074-7	$75.00
			Paperback	0-19-963075-5	$50.00
66.	Clinical Immunology	Gooi, H.C. & Chapel, H. (Eds)	Spiralbound hardback	0-19-963086-0	$69.95
			Paperback	0-19-963087-9	$50.00
65.	Receptor Biochemistry	Hulme, E.C. (Ed)	Paperback	0-19-963093-3	$50.00
64.	Gel Electrophoresis of Nucleic Acids (2/e)	Rickwood, D. & Hames, B.D. (Eds)	Spiralbound hardback	0-19-963082-8	$75.00
			Paperback	0-19-963083-6	$50.00
63.	Animal Virus Pathogenesis	Oldstone, M.B.A. (Ed)	Spiralbound hardback	0-19-963100-X	$68.00
			Paperback	0-19-963101-8	$40.00
62.	Flow Cytometry	Ormerod, M.G. (Ed)	Paperback	0-19-963053-4	$50.00
61.	Radioisotopes in Biology	Slater, R.J. (Ed)	Spiralbound hardback	0-19-963080-1	$75.00
			Paperback	0-19-963081-X	$45.00
60.	Biosensors	Cass, A.E.G. (Ed)	Spiralbound hardback	0-19-963046-1	$65.00
			Paperback	0-19-963047-X	$43.00
59.	Ribosomes and Protein Synthesis	Spedding, G. (Ed)	Spiralbound hardback	0-19-963104-2	$75.00
			Paperback	0-19-963105-0	$45.00
58.	Liposomes	New, R.R.C. (Ed)	Spiralbound hardback	0-19-963076-3	$70.00
			Paperback	0-19-963077-1	$45.00
57.	Fermentation	McNeil, B. & Harvey, L.M. (Eds)	Spiralbound hardback	0-19-963044-5	$65.00
			Paperback	0-19-963045-3	$39.00
56.	Protein Purification Applications	Harris, E.L.V. & Angal, S. (Eds)	Spiralbound hardback	0-19-963022-4	$54.00
			Paperback	0-19-963023-2	$36.00
55.	Nucleic Acids Sequencing	Howe, C.J. & Ward, E.S. (Eds)	Spiralbound hardback	0-19-963056-9	$59.00
			Paperback	0-19-963057-7	$38.00
54.	Protein Purification Methods	Harris, E.L.V. & Angal, S. (Eds)	Spiralbound hardback	0-19-963002-X	$60.00
			Paperback	0-19-963003-8	$40.00
53.	Solid Phase Peptide Synthesis	Atherton, E. & Sheppard, R.C.	Spiralbound hardback	0-19-963066-6	$58.00
			Paperback	0-19-963067-4	$39.95
52.	Medical Bacteriology	Hawkey, P.M. & Lewis, D.A. (Eds)	Paperback	0-19-963009-7	$50.00
51.	Proteolytic Enzymes	Beynon, R.J. & Bond, J.S. (Eds)	Spiralbound hardback	0-19-963058-5	$60.00
			Paperback	0-19-963059-3	$39.00
50.	Medical Mycology	Evans, E.G.V. & Richardson, M.D. (Eds)	Spiralbound hardback	0-19-963010-0	$69.95
			Paperback	0-19-963011-9	$50.00
49.	Computers in Microbiology	Bryant, T.N. & Wimpenny, J.W.T. (Eds)	Paperback	0-19-963015-1	$40.00
48.	Protein Sequencing	Findlay, J.B.C. & Geisow, M.J. (Eds)	Spiralbound hardback	0-19-963012-7	$56.00
			Paperback	0-19-963013-5	$38.00
47.	Cell Growth and Division	Baserga, R. (Ed)	Spiralbound hardback	0-19-963026-7	$62.00
			Paperback	0-19-963027-5	$38.00
46.	Protein Function	Creighton, T.E. (Ed)	Spiralbound hardback	0-19-963006-2	$65.00
			Paperback	0-19-963007-0	$45.00
45.	Protein Structure	Creighton, T.E. (Ed)	Spiralbound hardback	0-19-963000-3	$65.00
			Paperback	0-19-963001-1	$45.00
44.	Antibodies: Volume II	Catty, D. (Ed)	Spiralbound hardback	0-19-963018-6	$58.00
			Paperback	0-19-963019-4	$39.00
43.	HPLC of Macromolecules	Oliver, R.W.A. (Ed)	Spiralbound hardback	0-19-963020-8	$54.00
			Paperback	0-19-963021-6	$45.00
42.	Light Microscopy in Biology	Lacey, A.J. (Ed)	Spiralbound hardback	0-19-963036-4	$62.00
			Paperback	0-19-963037-2	$38.00
41.	Plant Molecular Biology	Shaw, C.H. (Ed)	Paperback	1-85221-056-7	$38.00
40.	Microcomputers in Physiology	Fraser, P.J. (Ed)	Spiralbound hardback	1-85221-129-6	$54.00
			Paperback	1-85221-130-X	$36.00
39.	Genome Analysis	Davies, K.E. (Ed)	Spiralbound hardback	1-85221-109-1	$54.00
			Paperback	1-85221-110-5	$36.00
38.	Antibodies: Volume I	Catty, D. (Ed)	Paperback	0-947946-85-3	$38.00
37.	Yeast	Campbell, I. & Duffus, J.H. (Eds)	Paperback	0-947946-79-9	$36.00

36.	**Mammalian Development** Monk, M. (Ed)			
......	Hardback	1-85221-030-3	**$60.00**	
......	Paperback	1-85221-029-X	**$45.00**	
35.	**Lymphocytes** Klaus, G.G.B. (Ed)			
......	Hardback	1-85221-018-4	**$54.00**	
34.	**Lymphokines and Interferons** Clemens, M.J., Morris, A.G. & others (Eds)			
......	Paperback	1-85221-035-4	**$44.00**	
33.	**Mitochondria** Darley-Usmar, V.M., Rickwood, D. & others (Eds)			
......	Hardback	1-85221-034-6	**$65.00**	
......	Paperback	1-85221-033-8	**$45.00**	
32.	**Prostaglandins and Related Substances** Benedetto, C., McDonald-Gibson, R.G. & others (Eds)			
......	Hardback	1-85221-032-X	**$58.00**	
......	Paperback	1-85221-031-1	**$38.00**	
31.	**DNA Cloning: Volume III** Glover, D.M. (Ed)			
......	Hardback	1-85221-049-4	**$56.00**	
......	Paperback	1-85221-048-6	**$36.00**	
30.	**Steroid Hormones** Green, B. & Leake, R.E. (Eds)			
......	Paperback	0-947946-53-5	**$40.00**	
29.	**Neurochemistry** Turner, A.J. & Bachelard, H.S. (Eds)			
......	Hardback	1-85221-028-1	**$56.00**	
......	Paperback	1-85221-027-3	**$36.00**	
28.	**Biological Membranes** Findlay, J.B.C. & Evans, W.H. (Eds)			
......	Hardback	0-947946-84-5	**$54.00**	
......	Paperback	0-947946-83-7	**$36.00**	
27.	**Nucleic Acid and Protein Sequence Analysis** Bishop, M.J. & Rawlings, C.J. (Eds)			
......	Hardback	1-85221-007-9	**$66.00**	
......	Paperback	1-85221-006-0	**$44.00**	
26.	**Electron Microscopy in Molecular Biology** Sommerville, J. & Scheer, U. (Eds)			
......	Hardback	0-947946-64-0	**$54.00**	
......	Paperback	0-947946-54-3	**$40.00**	
24.	**Spectrophotometry and Spectrofluorimetry** Harris, D.A. & Bashford, C.L. (Eds)			
......	Hardback	0-947946-69-1	**$56.00**	
......	Paperback	0-947946-46-2	**$39.95**	
23.	**Plasmids** Hardy, K.G. (Ed)			
......	Paperback	0-947946-81-0	**$36.00**	
22.	**Biochemical Toxicology** Snell, K. & Mullock, B. (Eds)			
......	Paperback	0-947946-52-7	**$40.00**	
19.	**Drosophila** Roberts, D.B. (Ed)			
......	Hardback	0-947946-66-7	**$67.50**	
......	Paperback	0-947946-45-4	**$46.00**	
17.	**Photosynthesis: Energy Transduction** Hipkins, M.F. & Baker, N.R. (Eds)			
......	Hardback	0-947946-63-2	**$54.00**	
......	Paperback	0-947946-51-9	**$36.00**	
16.	**Human Genetic Diseases** Davies, K.E. (Ed)			
......	Hardback	0-947946-76-4	**$60.00**	
......	Paperback	0-947946-75-0	**$34.00**	
14.	**Nucleic Acid Hybridisation** Hames, B.D. & Higgins, S.J. (Eds)			
......	Hardback	0-947946-61-6	**$60.00**	
......	Paperback	0-947946-23-3	**$36.00**	
12.	**Plant Cell Culture** Dixon, R.A. (Ed)			
......	Paperback	0-947946-22-5	**$36.00**	
11a.	**DNA Cloning: Volume I** Glover, D.M. (Ed)			
......	Paperback	0-947946-18-7	**$36.00**	
11b.	**DNA Cloning: Volume II** Glover, D.M. (Ed)			
......	Paperback	0-947946-19-5	**$36.00**	
10.	**Virology** Mahy, B.W.J. (Ed)			
......	Paperback	0-904147-78-9	**$40.00**	
9.	**Affinity Chromatography** Dean, P.D.G., Johnson, W.S. & others (Eds)			
......	Paperback	0-904147-71-1	**$36.00**	
7.	**Microcomputers in Biology** Ireland, C.R. & Long, S.P. (Eds)			
......	Paperback	0-904147-57-6	**$36.00**	
6.	**Oligonucleotide Synthesis** Gait, M.J. (Ed)			
......	Paperback	0-904147-74-6	**$38.00**	
5.	**Transcription and Translation** Hames, B.D. & Higgins, S.J. (Eds)			
......	Paperback	0-904147-52-5	**$38.00**	
3.	**Iodinated Density Gradient Media** Rickwood, D. (Ed)			
......	Paperback	0-904147-51-7	**$36.00**	

Sets

Essential Molecular Biology: 2 vol set Brown, T.A. (Ed)			
......	Spiralbound hardback	0-19-963114-X	**$118.00**
......	Paperback	0-19-963115-8	**$78.00**
Antibodies: 2 vol set Catty, D. (Ed)			
......	Paperback	0-19-963063-1	**$70.00**
Cellular and Molecular Neurobiology: 2 vol set Chad, J. & Wheal, H. (Eds)			
......	Spiralbound hardback	0-19-963255-3	**$133.00**
......	Paperback	0-19-963254-5	**$79.00**
Protein Structure and Protein Function: 2 vol set Creighton, T.E. (Ed)			
......	Spiralbound hardback	0-19-963064-X	**$114.00**
......	Paperback	0-19-963065-8	**$80.00**
DNA Cloning: 2 vol set Glover, D.M. (Ed)			
......	Paperback	1-85221-069-9	**$92.00**
Molecular Plant Pathology: 2 vol set Gurr, S.J., McPherson, M.J. & others (Eds)			
......	Spiralbound hardback	0-19-963354-1	**$110.00**
......	Paperback	0-19-963353-3	**$75.00**
Protein Purification Methods, and Protein Purification Applications: 2 vol set Harris, E.L.V. & Angal, S. (Eds)			
......	Spiralbound hardback	0-19-963048-8	**$98.00**
......	Paperback	0-19-963049-6	**$68.00**
Diagnostic Molecular Pathology: 2 vol set Herrington, C.S. & McGee, J. O'D. (Eds)			
......	Spiralbound hardback	0-19-963241-3	**$105.00**
......	Paperback	0-19-963240-5	**$69.00**
Receptor Biochemistry; Receptor-Effector Coupling; Receptor-Ligand Interactions: 3 vol set Hulme, E.C. (Ed)			
......	Paperback	0-19-963097-6	**$130.00**
Human Cytogenetics: (2/e): 2 vol set Rooney, D.E. & Czepulkowski, B.H. (Eds)			
......	Hardback	0-19-963314-2	**$130.00**
......	Paperback	0-19-963313-4	**$90.00**
Peptide Hormone Secretion/Peptide Hormone Action: 2 vol set Siddle, K. & Hutton, J.C. (Eds)			
......	Spiralbound hardback	0-19-963072-0	**$135.00**
......	Paperback	0-19-963073-9	**$90.00**

ORDER FORM for USA and Canada

Qty	ISBN	Author	Title	Amount
			S&H	
	CA and NC residents add appropriate sales tax			
			TOTAL	

Please add shipping and handling: US $2.50 for first book, (US $1.00 each book thereafter)

Name ....

Address ....

....

.... Zip ....

[ ] Please charge $ .... to my credit card
Mastercard/VISA/American Express (circle appropriate card)

Acct. .... Expiry date ....

Signature ....

Credit card account address if different from above:

....

.... Zip ....

[ ] I enclose a cheque for US $............

Mail orders to: Order Dept. Oxford University Press, 2001 Evans Road, Cary, NC 27513